Advances in Geographic Information Science

Series Editors
Shivanand Balram, Simon Fraser University, Burnaby, BC, Canada
Suzana Dragicevic, Burnaby, BC, Canada

The ***Advances in Geographic Information Science*** series aims to: present current and emerging innovations in GIScience; describe new and robust GIScience methods for use in transdisciplinary problem solving and decision making contexts, illustrate GIScience case studies for use in teaching and training situations, analyze GIScience tools at the forefront of scientific research, and examine the future of GIScience in an expanding knowledge-based economy. The scope of the series is broad and will encompass work from all subject disciplines that develop and use an explicitly spatial perspective for analysis and problem-solving.

Yan Ma • Zhenjiang Shen

Strategic Spatial Planning Support System for Sustainable Development

Agent-Based Modelling and Simulation

Springer

Yan Ma
School of Architecture and Urban-Rural Planning, Joint SPSD Lab
Fuzhou University
Fuzhou, China

Zhenjiang Shen
Joint SPSD Lab
Kanazawa University
Kanazawa, Japan

ISSN 1867-2434 ISSN 1867-2442 (electronic)
Advances in Geographic Information Science
ISBN 978-3-031-07542-1 ISBN 978-3-031-07543-8 (eBook)
https://doi.org/10.1007/978-3-031-07543-8

This Springer imprint is published by the registered company Springer Nature Switzerland AG
The registered company address is: Gewerbestrasse 11, 6330 Cham, Switzerland

Preface

Since planning measures have been proved to be effective in urban governance, our focuses on urban studies gradually expanded to how to make a good planning policy. While, it is a dialectical movement of cognition to evaluate on the planning policies good or not. It makes the decision-making process on such policies a scientific issue. To find a solution to this problem, different kinds of urban study models had been developed and employed in past decades. If we make a research review in this field, it could be found that the research methods experienced a developing process went that from macro to micro, from aggregated to disaggregated, and from static to dynamic. Although research methods may be different, the purposes of such researches are similar, which evaluate or forecast the effects of targeted planning policies by building models. Accompanying the developments of decision-making support in urban planning, can we find or build a research format in this area?

This book will introduce how to develop planning support tools by combining policy decision-making process with the mechanism of urban spatial or functional development regarding the planning policies on a strategic level. We will concentrate on visualizing and forecasting the macro phenomenon of policy effects through revealing the discrete, micro human behaviors around the urban functions of dwelling, recreation, working, and transportation. The models are developed on constraint cellular automata (CCA) approach or multi-agent system (MAS) approach and covered the urban developmental stages of urbanization, suburbanization, and urban decline.

For a long time, researchers kept their sights on how to forecast or visualize the effectiveness of a certain policy. We may think the forecasting or visualization of a policy is exactly one kind of planning support, which is enough to support the policy makers to make decisions on their argued policies. However, regarding decision-making on whether a policy is good or not, there should be a constraint for the forecasting or visualization results. This is also one important point of the decision-making process we talked in this book. Usually, environmental impacts could be evaluated for such constraints, especially for the strategic plans which would result in urban spatial changes. While for some public policies such as development of public facility, residential allowance, or price adjustment of some public goods, the satisfaction of residents on the argued policies or designing schemes may be a suitable indicator candidate for the constraint we mentioned above.

In this book, we will use the environmental impacts as a constraint to compare different planning schemes. Considering the strategy of carbon peaking and carbon neutrality in China, today, we developed a simulation tool to evaluate the low carbon degree of different planning schemes regarding inter-city transportation and land uses derived from urban master plan. Meanwhile, we also compared the differences between individual behaviors and agent-based behaviors. We introduced how to employ agent-based models (ABMs) for planning support through several empirical studies.

Although we think there are still some constraints to employing ABMs for planning practice, such as the seldom-validated decision-making rules for agents, the demand of accurate discrete data sources for defining agent behaviors, or computing power of simulation platforms, we still believe that the research improvements on big data and artificial intelligence will bring ABMs a practical future.

Fuzhou City, Fujian Province, China Yan Ma
December 2021

Acknowledgments

Because of the COVID-19 pandemic, I could not get back to my office for a semester to continue with the writing process, and this had negative impact on the book's schedule. But the team from Springer Nature never gave up. In this case, I wish to thank the editors, book coordinators, and staffs who continuingly contributed to the publishing process of this book during the hard time. Your contributions make the work more meaningful.

In the meantime, I would like to thank Prof. REN Yunying and other research team members at Xi'an University of Architecture and Technology who supplied us the data sources for Chap. 8; Prof. XU Hui, China Academy of Urban Planning and Design, who supplied the data sources of Jiangxi Province, China, for the research experiments of the second case study in Chap. 3; our master's students Ms. LIN Hanyu and Mr. LIN Yifei, who prepared the material for Chap. 6; Mr. XUE Feng cooperated with me to conduct the experiments of the second case study in Chap. 3; Ms. ZHANG Huanli cooperated with me to conduct the experiments in Chap. 8 and checked through the format of the book; and Ms. TANG Ningjun checked through the figures in the book.

Parts of this book have appeared before. In Chap. 3, some material of the first case is taken from a paper published at the 18th International Conference on GeoInformatics, 2010. In Chap. 4, the parameters of the simulation model are from a paper published in the *Journal of Artificial Societies and Social Simulation* (03.06.2013). The model design in Chap. 5 was introduced in a paper published at the Annual Conference of the City Planning Institute of Japan (11.18, 2006).

Contents

Chapter 1
Introduction to Agent-Based Modeling and Its Application to Policy Decision-Making

1.1 Introduction

In recent decades, the widely accepted concept of sustainable development has made policy approach for realizing sustainable urban development a hot concern in urban planning and design.

While these years urbanization is taking place at an unprecedented pace in developing countries. There is no doubt that without reasonable management and intervention, an excessive urbanization will result to excessive aggregation of urban population, deterioration of urban landscape, growth of urban crime rate, and also regional water shortage. Thus, when making a decision on the policies that would lead to urbanization, planners or government decision makers should find a decision-making way that can take these possibilities into account and forecast the possible influences of the policies. In the contrast, deurbanization is taking place in some developed counties, which resulted to low utilization rate of city facilities, downtown decline and also private car dependency. Regarding to urban decline, researchers are trying to use policy approach to make their cities more compact and their downtown more active. The negative results of these two urban development stages both make urban sustainable development a challenge not only for researchers but also for the decision makers who make decision on urban development related policies.

This book targets to supply planners and government decision makers who make decision on urban development related policies with planning support tools. These tools are mostly agent-based models and supposed to be able to simulate and visualize household behaviors in urban area affected by these planning policies. These behaviors include population aggregation and household residential mobilities during urban growth, shopping behaviors in different urban areas and meanwhile, the relevant public demand for community facilities and their location choices, household commuting choices. Moreover, we also focus on the environmental influences

Y. Ma, Z. Shen, *Strategic Spatial Planning Support System for Sustainable Development*, Advances in Geographic Information Science,
https://doi.org/10.1007/978-3-031-07543-8_1

resulted from urban development, such as urban water management. Different from existed modeling researches, the models introduced in this book are targeted to special planning policies and supposed to be useful in decision-making support. Through simulation on urban growth, population aggregation and household residential mobility, location choices of community facilities, household shopping behaviors and commuting choices, we suppose that government decision-makers would be able to more easily make decisions on relevant planning policies.

1.2 Models for Analyzing Urban Spatial Strategic Plans

1.2.1 Models for Analyzing Urban Policies

As showed by European experience in urban development, generally there are three stages for an urban development (Klaassen and Paelinck 1979), urbanization, sub-urbanization and urban decline. During different stages, there are kinds of policies supporting government to manage on urban development issues. These relevant policies cover a very wide range of aspects and follow a top-down relationship. Such as spatial plan, land use plan and urban plan, etc.

Regarding how to examine the impacts of such kind policies, the related literatures are vast. The statistical analysis has been proved as an effective approach in describing the impact of the planning policies. For example, Mitani and Yamanaka (1996) used build-up speed, land price and local street network as evaluation indicators to analyze the impact of the development ordinance and local area fund system on the district urbanization in Komatsujima City of Japan. Researchers often use regression analysis to study Land Use/Cover Change (LUCC) phenomena due to its ability to test theoretical assumptions, rank relative factors, and yield rigorous hypotheses test (Yu and Ho 2006). But LUCC can be resulted from the complex interaction of social, ecological and geophysical processes, cannot simply be fully explained and forecast in the regression model by a set of carefully selected independent variables. This approach is hereby weak to uncover the underlying processes that give rise to such patterns and also hardly able to forecast what will happen if an alternative policy were performed.

Due to forecasting, there have been researchers using elaborate economic models to evaluate or forecast the impact urban planning policies (Kazuki Nakamura et al. 2004). However considering the complexities of the urban system, equilibrium conditions are not easily reached. And relationship identified using economic models does not readily revealed the dynamic process leading to the relationship. Some significant advances have been made through applying traditional policy impact analysis model in geographic information system (GIS) environment. As a useful method to represent, analysis and display the spatial data, Geographical Information System (GIS) could be used to reflect the spatial change in an urban area through comparing and analysis the spatial data collected in different stage by remote sensing.

However, the approaches above just as pointed by Batty (2005), they do not revolve around their equilibrating structure and aggregative nature for addressing the concerns of contemporary planning and policy analysis.

1.2.2 MAS and CCA for Urban Modeling

As argued that GIS failed to capture the dynamic process of urban development. While many works have demonstrated that Cellular Automata (CA) approach has great capability to simulate spatial process of urban development (Li and Yeh 2000; Shen and Kawakami 2008). While, urban development is a complex process that would be influenced by many macro-scale driving forces. The pure CA model is inadequate to taking these macro-scale political, economic and cultural driving forces of urban growth into account (White and Engelen 1997). To improve a CA model to represent the complexity spatial process of urban development, the constrained CA model has been applied for simulating urban boundary change influenced by local, regional and global sustainability constraints (Li and Yeh 2000).

Despite of advantages in spatial presentation, CA approach also has its own drawbacks. CA model likes arrangements of individual automata over tessellated space, where an individual automaton is influenced by automata in neighboring cells (Torrens and Benenson 2005). It results in the shortage of CA approach that cells cannot move and do their own decisions (Torrens 2007). Therefore, although CA-based models have strong ability to represent and simulate spatial processes of urban growth underlying spatial decisions related to neighborhood interactions (Batty et al. 1997), ABM can much more refer to the individual decisions.

As reported previously (Jager and Mosler 2007), ABM is expected to contribute to the exploration of the effectiveness of policy measures in complex environments through behavior-environment interactions. A number of studies have used an ABM to assess future socio-ecological consequences resulting from land-use policies (Lee et al. 2010) and other studies have focused on the use of multi-agent simulation for policy development (Berger et al. 2006). ABMs are widely used for residential simulation from the viewpoint of economics with respect to real estate, in which housing prices are an important factor. Based on the assumption that gentrification occurs because capital returns to the inner city and creates opportunities for residential relocation, Diappi and Bolchi (2006) presented a dynamic model developed on a Netlogo platform using a multi-agent/cellular automata system approach. By describing the relocations of households, researchers have to some extent addressed the housing market processes and price formation (Ettema 2011; Dawn and Filatova 2008). In particular, Moeckel simulated households as evaluating individual dwellings until eventually finding and accepting a dwelling that offers a significant improvement over their current dwelling (Moeckel et al. 2003).

1.2.3 Conclusion for Literature Review

This review of the recent literature regarding tools and techniques to support policy analysis indicates that while traditional approaches have been extensively and successfully used in supporting urban planning practice, they have limits in studying the underlying process that give rise to urban macro patterns and in situations where decision-making are made under conditions of deep uncertainty. Nevertheless, no single technique can capture the richness of an urban system, and it makes sense to use a diversity of tools to build virtual environments, particularly when they are to be employed as planning support tools (Torrens 2007).

Comparing with these traditional approaches, CCA approach and MAS approach are capable to better serve these situations. Moreover, the combined CCA and MAS modeling can be very suitable for simulation on urban spatial process and individual behaviours in urban space. Because MAS can evaluate the impacts of policy changes at the level of the decision-making units such as government, facility owners and households rather than focusing on the aggregate information of groups of individuals to predict policy impacts, while at the same time estimates of aggregate outcomes can still be derived by summing up individual predictions. Flexibility in designing new action rules and environment constraints in a simulation allows researchers to test alternative policy scenarios.

Despite the well-documented MAS and CCA for modeling, surprisingly there are few researches have addressed impacts of urban growth policy or urban center regeneration policy. In light of this gap in the literature, our research seeks to explore MAS to simulate the possible impact of different policies within different urban development stages.

1.3 What Is Agent-Based Models

1.3.1 The Definition of ABMs

In a long time, the urban issues in developing countries are focusing on urban growth, population aggregation, environmental pollution prevention, etc. Researchers manage to use planning policies to make urban development under their control, for this purpose we try to calculate the population scale, the demand of urban development land and public facilities by using models and methods as detail as we can. In Chinese planning management system, planners usually use the population growing rate in the past 10–20 years to forecast the population scale in the coming future and a factor named per thousand person's index is widely used in planning practice and management to calculate households' demand for different development land and facilities. In this kind method the personal characteristics and status of households or population are ignored, we thus call this kind model aggregated model. Even there are rules and laws among the entities, they are relatively few and are mostly focusing on global phenomena. The applications of aggregated models in policy analysis thus are good at reflecting the top-down process.

Due to the complexity of urban system, models used in planning analysis could be characterized by the bottom-up process (Epstein et al. 2011). Phenomena thus would be studied as the product of multiple interactions among discrete and independent entities, for example, studying the policy impacts of a determined residential policy would focusing on the residential location choices by each household living in the city and their interaction (see in Chap. 4). In such kind case, the individual behavior is nether collected by historical data nor simple self-organized, it is discrete, adaptive and interactive. The Agent based modeling (ABM) is one of the representative methods. Agent based modeling (ABM) is an emerging approach to modeling complex processes and phenomena in social science in recent years. Also named as multi-agent system (MAS), agent-based simulation (ABS). There are researches took "cellular automata"as the first prototype of the interactive behaviors through grids (Von Neumann 1951; Chen 2012). Although the cellular automata can simulate interactions between cells, the cells cannot make decisions by themselves.

Basing on the illustration above, it could be concluded that an ABM is a disaggregated model, by which the phenomenon is studied as a result of agents' behaviors and interactions. The decision-making of each agent is seemed independent and, in the meantime, affected by interactive process between other agents. This character means the agents behaviors are not always stay the same, they are changeable and adaptive, so that, reflecting the uncertainty of a phenomenon through bottom-up. An ABM thus is usually composed by three parts, the agents who play different behaviors in the model, the environment where the agents behaves and last, the interactions between agents for them learn and make final decisions on their behaviors. In the next subsection I will further introduce how to build an ABM.

1.3.2 How to Build a ABM

1.3.2.1 The Agents

As the basic unit of ABMs, the agent should be defined and explained at the first place when an ABM is going to be developed. Then a question appears that what an agent is. Actually there are various definitions about agent. Some researchers think an individual agent is "just something that perceives and acts" (Russell and Norvig 1995, p. 7), this definition shows a basic character of agent that it can act. While this is still not enough, a more rigorous definition of agent shows that "Autonomous agents are computational systems that inhabit some complex dynamic environment, sense and act autonomously in this environment, and by doing so realize a set of goals or tasks for which they are designed" (Maes 1995). We will find some key points from this definition that an agent is not only be able to act, but also autonomous, it means agents can make decisions by themselves. Furthermore, they make decisions in some complex dynamic environment. Since then, the characters of agent could be summarized as social ability, reactivity and proactiveness.

Thus, agent is an actor in a complex system, no matter this agent is an animal or humanity, the essence of an agent is that it can make decisions on its own actions independently. In the meantime, its actions would be influenced by other agents in the system and hereby, shows a character of flexible, adaptive and interactive. Due to different research purpose, the agent will de defined with different attributes. These attributes are the inputs of the simulation system from external, and will be output after internal process. Within the internal process, the agents will take actions and interact with other agents and the environment (Fig. 1.1). For an example, in a research topic of residential location choice, whose behaviors affect the residential location choices will be defined as agents in the model, if we do not consider the government's management in the process, the human or household who make location choices for residence will be seemed as one kind of agent. The agent should also be defined with special social attributes just the same as them in the real social system. Figure 1.2 shows an agent system in a residential relocation model, in which agent stands for households in a society, they have different income, special locations, household age, sexual and household members, just like what they are in a real society. The agents will make location choices on each land use plot through the interactions between each other and the environment they living, and finally produces a residential location pattern in the environment.

If your research concern is about ecological issue, the agents could be animals or insects too. Just as a model developed by the MIT Media Lab, it focused on the colony of ants forages for food (Resnick 1994). In the model, each ant lives in a determined environment, it has own nest and finds food every day. When ant finds food, it will bring the food to nest. In order to come back again, the ant will drop a chemical on the patch (the spatial unit in the simulation model). In this case, other

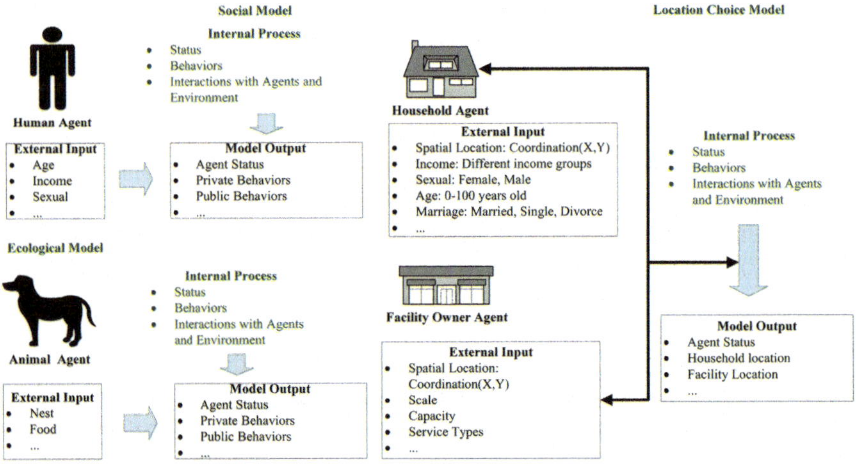

Fig. 1.1 Characters and acts of agents in ABMs (as an example of location choice model)

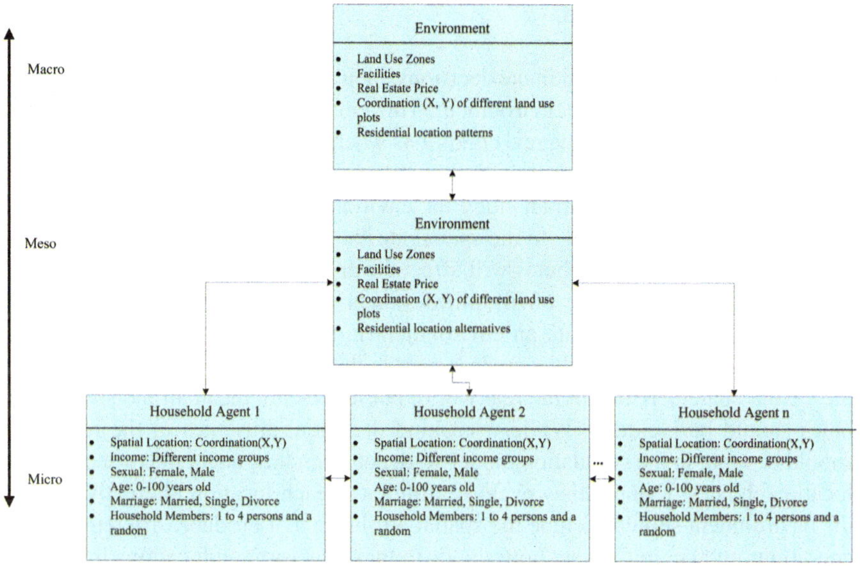

Fig. 1.2 an example of agents' attributes and hierarchical structure of the ABM for household residential location choice

ants will follow the chemical to find food on the same place. This process shows the interaction between ant agents. Comparing this case with the household location choice model, we will find although ants are different from humanity, they also make decisions on their behaviors independently. Thus, it will be the essential character to distinct the agent from general entities, such like the grid in a cellular automata model, although there is interactive process between grids, the grids are not suitable to be defined as agents due to the grids are not autonomous.

Accompanying with the development of artificial intelligence, the agents' definition is going to be extended too. An agent is not only autonomous but also intelligent. The intelligence of an agent gives it a new ability, which is learning from surroundings. Through the interactions with other agents and the environment, one agent will make a decision on its behavior. It means agents' behaviors are not immutable, instead of, agents would probably adapt to the surroundings through learning and then change the behaviors it used to do. For example, in a commuting model, one agent is usually taking bus to work and this behavior has been lasting years. While due to a new residential development program, the bus he/she taking becomes very crowed and the waiting time till he/she get on the bus becomes longer too. The extended time makes he/she late for work. When this situation lasts several days, the agent gives up taking the bus he/she used to and choose another commuting mode correspondingly. Through re-create and predict the appearance of complex phenomena.

1.3.2.2 The Environment

No matter how does an agent make decisions, their behaviors and interactions should be carrying by a non-agent environment. This environment created in a computing system is usually to reflect the real conditions where the agents behaves in a complex system. Thus, in different research area the information and attributes of the environment will be different. In urban study the environment is usually created to reflect spatial conditions. Regarding to the spatial attributes of the environment, it depends on what kind of spatial attributes will affect the agent behaviors. In the case of residential location model, the environment stands for an urban area where all the households live. When create an environment in the computing system to accommodating household agents, the environment in the computing system should follow the spatial characteristics in the real world, but it does not mean all the spatial attributes should be included. While what kind of spatial attributes in the real world should be input in the simulation model? It depends on how agent make decisions on location choices. Usually they evaluate utilities of each location alternatives. Thus, the environment should include the spatial attributes that produce different location conditions and hereby, attract households to make decision on location choices.

As what has been showed in Fig. 1.2, the spatial attributes of environment show strong geographic characteristics. Thus, the environment could be linked to geographic information system (GIS). No matter in raster or vector format, the GIS dataset obtained from real world needs to be input to a simulation platform or software to create the environment. The environment of an ABM will have the basic geographic information such as the area, location of each grid or polygon, due to different purpose, multiple map layers could be used to reflect targeted spatial features. The availability of detailed dataset gained from GIS offers opportunities to create highly 'realistic' environments that agents can inhabit and interact with (Fig. 1.3).

1.3.2.3 Agent's Decision Rule

Agents' behaviors in an environment are autonomous, adaptive and uncertainty, but follow special rules. The rules in the simulation system are a set of commands that predefined in an ABM and probably contain two aspects of meaning. From one aspect, the rules guide agents make decisions on their behaviors. From the other aspect, these rules also influence on how an agent keep relationships with other agents and surrounding environment. It is obviously that agents cannot take actions in a model without of decision rules. Although important, it remains a significant challenge to set such rules for agents. In existing ABM researches, agents' decision rules are usually set by referring to published literatures, expert knowledge and data analysis. More accuracy the rules are, the more realistic the model could be.

The rules for agents making decision on their behaviors are basically follow a "if else" logic, if a predefined condition is met, the agent will take the related action, otherwise, choose the other one (Fig. 1.4). There could be one single rule or a set of

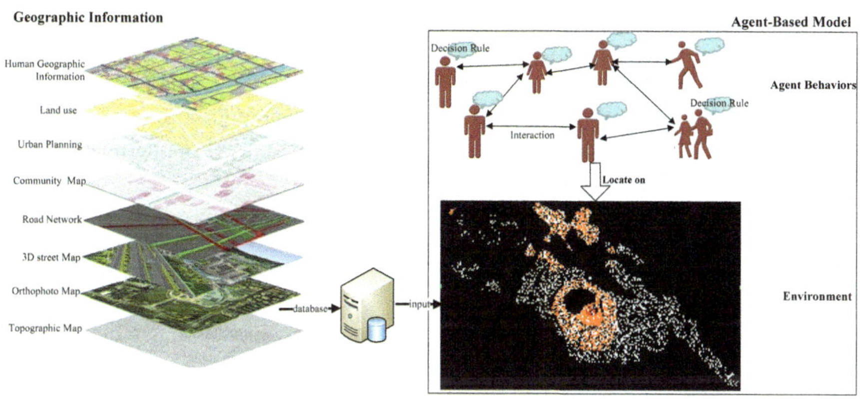

Fig. 1.3 Conceptual relationship between geographic information and environment in an ABM for accommodating agents' behaviors and interactions

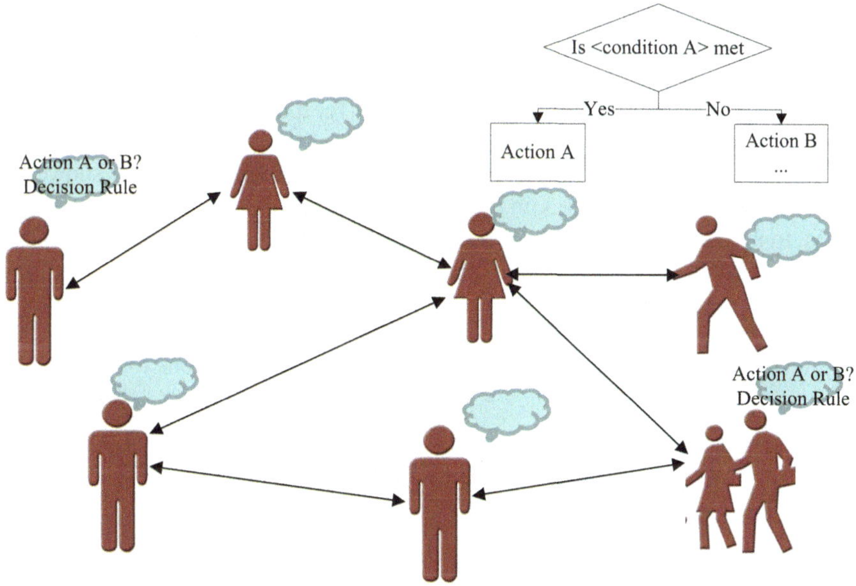

Fig. 1.4 the conceptual scheme of agents' behavior choices basing on decision rules in ABM

rules for an ABM, and all agents could follow one decision rule or different rules. Sometimes although different agents follow the same decision rule, the condition for them make a decision could be different. MA et al. (2013) created household agents that all follow one single rule that make residential location basing on desire to maximized utility. But the utility preference of each agent is different. In the ShopSim model (Shen et al. 2011), there are two kinds of agents, the shop owner and household respectively, thus they make decision choices basing on different decision rules. The similar cases are vast. Agents also learn to make decisions through evolutionary computation, and accompanying with the development of intelligence artificial this ability of agents are going to be strengthened in building an ABM. In the meantime, the agent decision rule is going to be more realistic by the increasing of multi-sourced big data, such as the spatial-temporal human behavior data.

1.3.2.4 Agents' Interaction

Although agents are making decisions and choices basing on designed decision rules, models with decision rules for individual behavior are not all agent-based. The core difference is whether there are interactions between individuals. If we take a half day's individual behavior as an example, we can research on individual' s behavior by questionnaire, field survey or use GPS to record individual's location at each observing time. Basing on these data and observations, we can build individual-based model through connecting the observed behaviors with individual's social economic attributes. If the survey or observation covers a relatively long period, it would be possible to recognize one's different behavior choices at the same time each day. It allows us to reveal the decision rules for the individual choosing behaviors in the model.

There are two types of individual-based models, X and Y respectively showed by Figs. 1.5 and 1.6. In model X individual A usually takes 1 hour's walk from 7:30 A.M. to 8:30 A.M after breakfast, but it depends on the weather. Thus, in the model a "if else" rule is designed for individual A making decision on taking walk or not. After taking walk, individual A usually take a rest outside till lunch time. The situation is similar to individual B and C, who make decision on his/her behaviors basing on determined rules in the model. The same situation happens to individual B

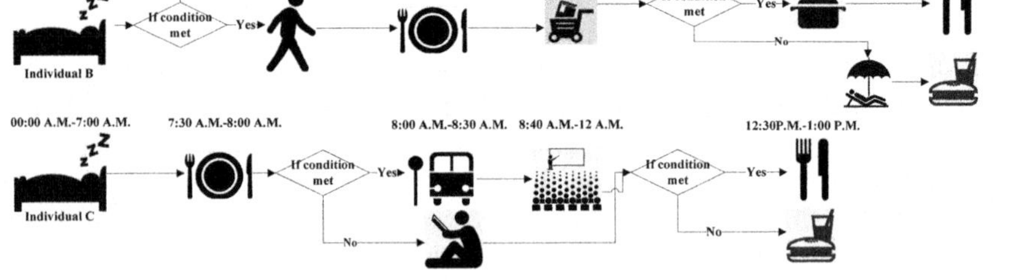

Fig. 1.5 a conceptual model X for reflecting a half day's individual behavior

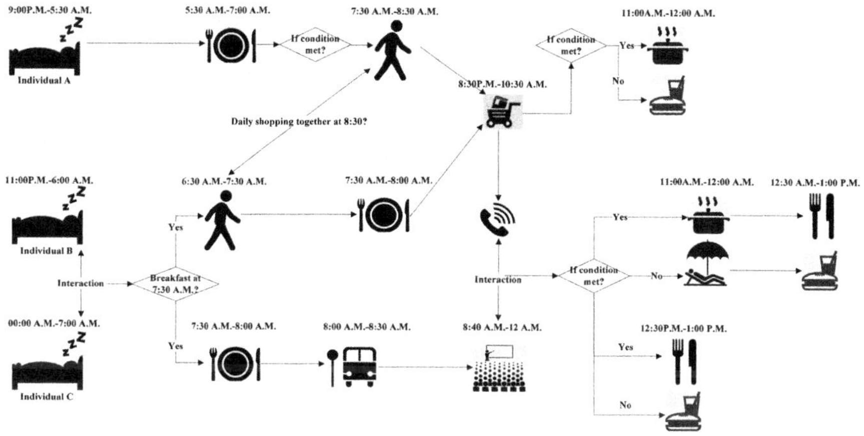

Fig. 1.6 A conceptual model Y for reflecting a half day's individual behavior

who needs to make a decision on his/her breakfast time and lunch cooking. These information together with the decision rule are usually revealed through statistical analysis, but in this conceptual model behaviors are recorded and researched as independent events. The similar logic is widely employed to explain the relationship between human spatial-temporal behavior and urban spatial structure, built-up environment or pollution exposure (Yu et al. 2018; Shen et al. 2020; Ma et al. 2020). It is able to explain or assess the current status of a research issue affected by human behavior very detail.

But model X is hard to be employ to forecasting the future, due to the choices and decision process in a real society would be much more complicated than what could be reflected in such an individual model. In an artificial society, individuals are not separated from each other, although individual A takes rest since 8:30 A.M.to 10:30 p.m., it does not mean he never change. If one day he/she met a neighbor who invite him/her to join daily shopping, individual A probably change his/her behavior at that time. The same situation also happens to individual B, as a common sense, human will have breakfast when they wake up, but in this case individual B's breakfast time depends on his/her family member individual C's waking up time. These uncertain situations are concerned and contained in model Y. Thus, in the individual-based model X the decision rule has not took interactions of individuals into account. Although individuals make decision independently, they are not separated. It is common that individuals will be influenced by other individuals during they make a decision. Just like individual B will interact with individual C to determine his/her schedule for next day's breakfast. In the meantime, after individual B's daily shopping, his/her choice on cooking or not also will be influenced by individual C. The rule is that if individual C came back home for lunch, individual B will cook, otherwise, he/she usually takes rest till 12:00 a.m. and takes some fast food for lunch.

Basing on this example, we can see that in an individual-based behavior model, if there are no interactions between each individual, the model is hard to reflect the uncertainty and complexity of artificial society. The individuals in such kind model is relatively weak in adaptive behaviors. Such as the probability of individual A join individual B's daily shopping concerned in model Y, will be ignored in model X. Thus, although individual-based, model X is one kind of micro model but not agent-based. The typical characteristic of ABM is that an agent's decision is making on the interactive process with other agents, during the interactive process, agents will learn and adapt to taking a more suitable action. The decision rule set for agents hereby should be able to reflect interactions.

1.4 Agent-Based Models for Decision Support of Urban Planning Policies

1.4.1 ABMs for Individual-Based Behavior Research

As introduced above, agent-based model is much more suitable in reflecting the uncertainty of individual behavior than other individual-based micro model. So that the existing researches by using ABM are abundant in quantity and wide in range. As a method by researching on macro patterns through micro-level agent behaviors, ABM is widely used in social science.

A typical example is the segregation model of Schelling (1971) implemented in Netlogo platform. In this model there are two types of agents, namely red and green. The red agents and green agents get along with one another. But each agent wants to live in a neighborhood, which is defined by Moor8 neibourhood (one cell's 8 surrounding cells are seemed as its neighbors, where a certain percentage of neighbors are the same color (here a certain percentage is chose). The rule for agents choosing location is that each red agent wants to live near at least some red agents and each green agent wants to live near at least some green agents. The simulation shows how these individual preferences effects on their location choices and finally produce a large-scale pattern. By adjusting the percentage of similar neighbors in a neighborhood from 30% (initial parameter) to 60%, the segregation pattern could be observed in Fig. 1.7. ABM is also built to research on how wealth distributions emerge in a society (Chinesta 2021) and how virus spreads in a society (Dhou and Cruzen 2021). Figure 1.8 shows the simulating results of wealth distribution and the spread of AIDS in Netlogo.

These examples show us the advantage of ABM in reflecting the multiple macro patterns through simulating micro behaviors. Although these examples are abstract or academic, it does not mean that ABMs are not suitable in other field except for academia. As proved by exist research, ABM is expected to contribute to policy decision-making through simulating the interactive process of individual behavior and policy intervention (Jager and Mosler 2007). Such as the assessment of future socio-ecological consequences resulting from land-use policies (Lee et al. 2010), forecasting the effectiveness of downtown regeneration policy by simulating

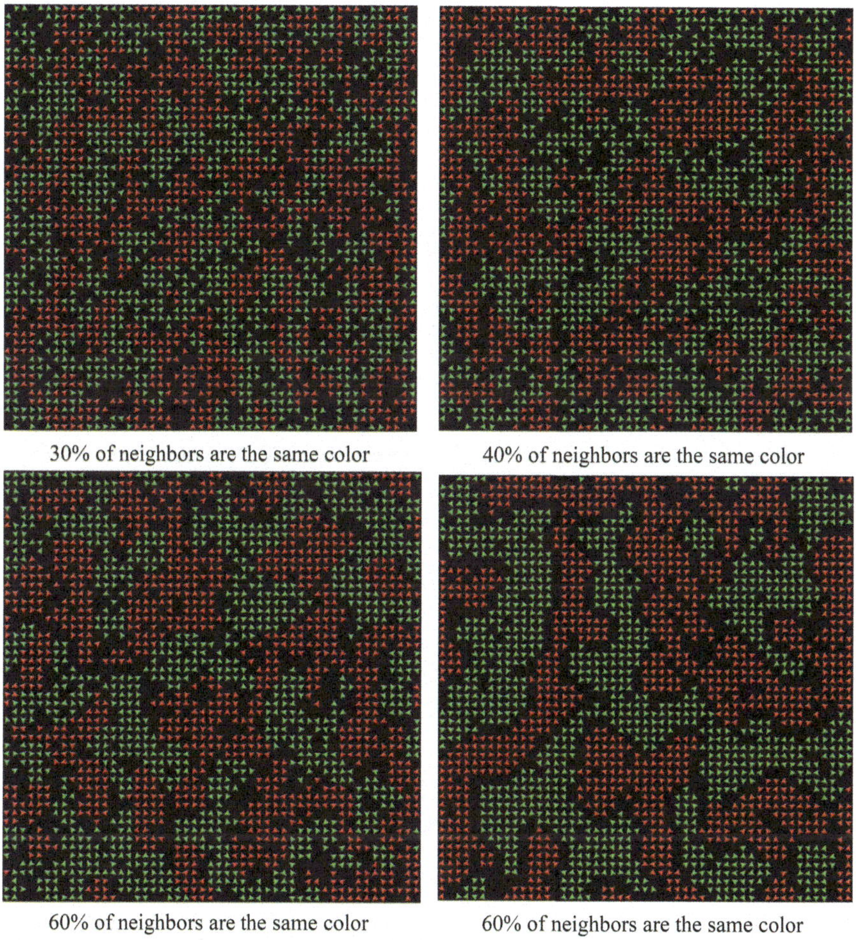

Fig. 1.7 the segregation pattern of agents with different percentage of neighbors in Netlogo Platform

household residential location choices (MA et al. 2013). Researchers also focuses on using ABM to represent the complexity of urban system, such as the UrbanSim model, which is also developed as ABM for simulating different actors and processes in the urban environment, such as residents, businesses, land developers, and transportation networks (Noth et al. 2003). ABMs are also widely employed to simulate consumer behaviors in the support of retail space design and planning (Shen et al. 2011; Bonabeau 2003). These examples and not only limited to them, prove the possibility of ABM in policy decision-making.

But if we trace back why we choose ABM to research on individual behavior rather than ordinary individual-based micro model, we need to return to the examples showed by Figs. 1.7 and 1.8. In these examples we can observe the process of

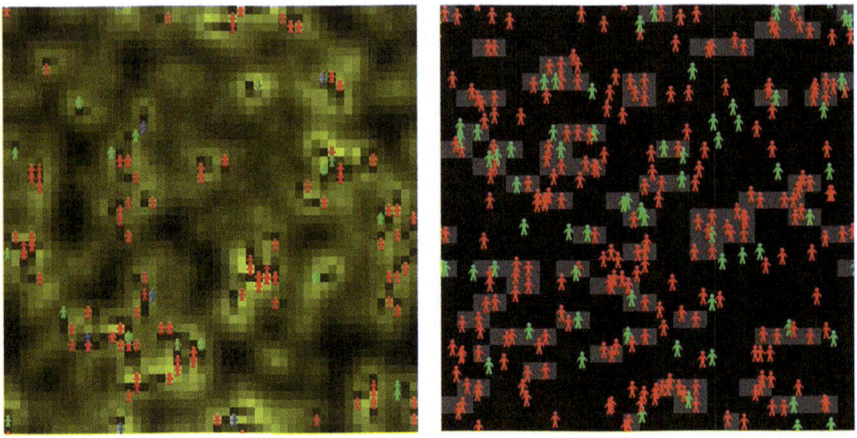

The result of wealth distribution model in Netlogo The result of AIDS spread model in Netlogo

Fig. 1.8 ABMs for simulating wealth distribution and AIDS spread in Netlogo

agents status changing (including segregation, wealth distribution and AIDS infection) at a certain time and under a predefined condition, but it is hard to be observed in reality. Because these processes can only be observed when they are undergoing so that the detailed chronology becomes impossible to reconstruct (Batty 2005). That is why we need ABM to research individual behaviors.

1.4.2 ABMs for Policy Decision-Making

The individual behavior in an ABM and related macro pattern just like the quantum motion in quantum mechanics, is taking place all the time and constantly in flux. What model developers need to do is to make the agent decision rule, interaction and living environment close to reality. From this point, if an ABM is developed to policy decision-making, policy making is indeed a kind of individual behavior of government officers. Although they make policies, the behavior is not isolated. During making a policy, at least two kinds of entities' behaviors need to be concluded in the model, which are government officers and stakeholders respectively. The former responsible for arguing a policy, and correspondingly the latter will be influenced by the policy and produce a feedback to the former. The government officers' decision-making will be affected by other stakeholders' feedback too. This process is a typical interaction process between individuals.

Therefore, when developing an ABM to support the decision-making of a targeted policy, the decision process should be developed and combined with agent behavior. The decision process contains two levels, the first level is to represent the interaction between government policy maker and stakeholders, the second level is to design the decision rule for government policy maker. Figure 1.9 shows the

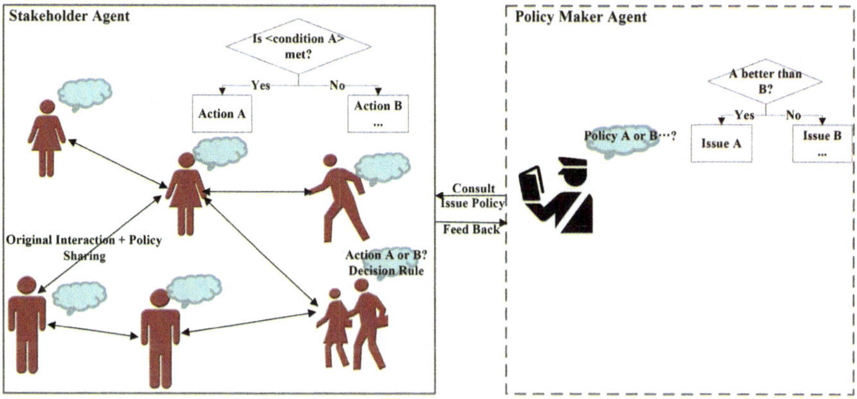

Fig. 1.9 the conceptualization of agent interaction during policy making in an artificial social system

conceptual process of interaction between policy maker and other stakeholders in an ABM. As shown in this figure we can see that when issue a policy government agent will consult with the stakeholders and make a decision on different policy schemes basing on the feedbacks.

Stakeholders' feedbacks will also contain two levels' information, on the first level it represents one individual's original policy attitude but on the second level it will also be influenced by other individuals by policy sharing, such as neighbors, family or friends. This process will effect on one individual's final decision on his/her behavior. For example, one local government issued an affordable house policy, there is one individual going to buy an apartment, will he/she use the policy? If his/her neighbors are very positive to the policy, he/she will receive positive influence too and vice versa.

Besides the interaction between government policy makers and stakeholders, the decision process of government agent also needs to be designed in the model. This process should follow the same process in real society. It is designed and showed as Fig. 1.10. The government agent needs to determine the concern issue before argue a policy and then, define the planning objectives. The former is usually basing on social management requirements and latter is interact with stakeholders. When planning objectives are clear, a scenario analysis is employed for making a decision. Because the policy attitudes and feedbacks of stakeholders will influence the effectiveness of policy implementation, so that government strategy developing will involve stakeholders. That is exactly what collaborative planning or design does in urban planning in reality. In our field we also can see that government will organize public hearing for a coming policy. The last step is policy implementation in the model and government agent will monitoring the policy implementation. When the policy has been implemented, stakeholders behaviors will be influenced by it and their decision rule and interactions will be influenced too. And finally a macro

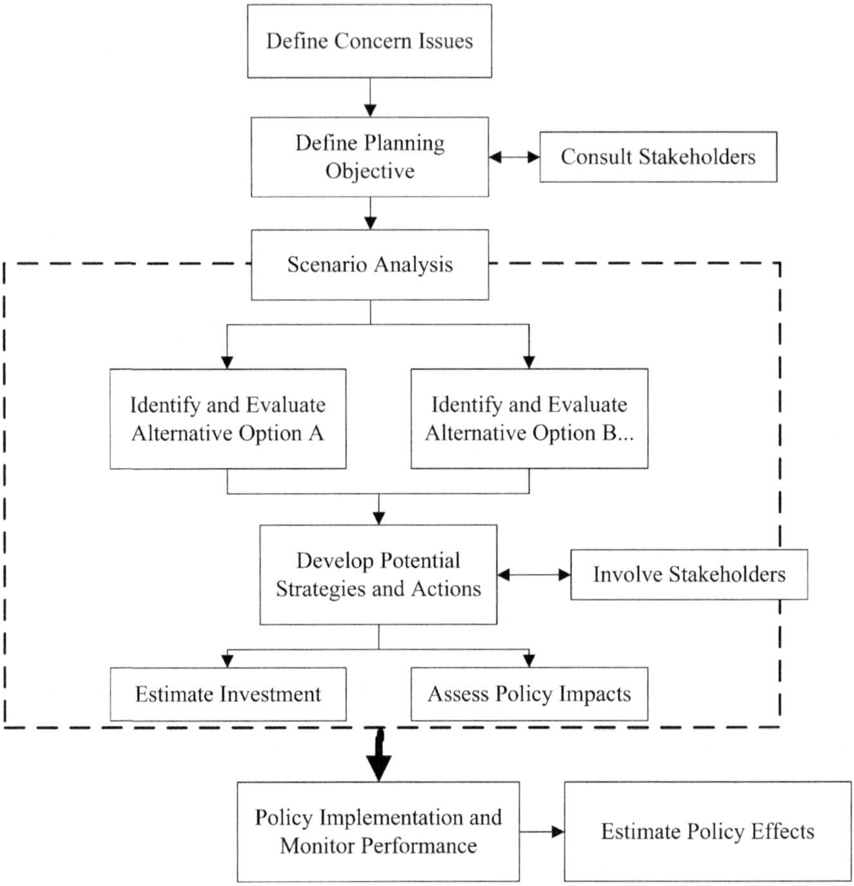

Fig. 1.10 The decision-making process in an artificial social system

pattern of the policy effects will gain from each stakeholder's individual behavior. this macro pattern could be referred to make a final decision.

1.5 Discussion

ABMs are developing with an unprecedented speed in academy and practice over the last decades. This benefits from the rapid development of computer science and the more and more friendly development environment. Even ABMs are good at reflecting the complex of individual behaviors, the decision rules of human behaviors and interactions between individuals are closely related to existing researches and micro data source. To some extent, the application of ABMs in practice is limited to data gaining. But the current development of new data science creates new

	Urban boundary change	Household mobility and location choices	Location choice of tertiary sector	Environmental impacts
Investigation contents	**Spatial data,** from statistic books, GIS and remote sensing data in case study area	**Housing data,** housing prices, household mobility in case study area within past 10-20 years	**Tertiary data,** industry and commercial locations in case study area within past 10-20 years	**Water/soil/ air/ solid waste/ noise (here water resource consumption has been took as an example),** industry inventory and the volume of their water consumption
	Policy, land use policy, land use control, urban plan, land use suitability, social economic development plan, etc.	**Housing policy,** residential promoting policy, affordable housing policy, etc.	**Policy,** development strategies	**Water supply capacity, waste water disposal capacity,** charge price of water supply, charge price of waste water disposal
Model development approach	Constraint Cellular Automata (CCA) approach	Multi-Agent System approach	Multi-Agent System approach	Multi-Agent System approach
Simulation model development	**Urban growth module**	**Household lifecycle stage module and household location choice module**	**Industry and commercial location choice module**	**Inventory based urban agents' water consumption module**

Fig. 1.11 the framework of urban development strategy modeling by using micro simulations

opportunity for ABM gaining multi-source individual-based data and hereby, reveal and verify the decision rule of individual.

Obviously, this is a very positive to whom is interested in ABM. For many applications that researchers seem individuals as the core of a complex social system, in which all the consequences are resulted from individual's decision-making over space and time. To research this process more detail, researcher build various simulation models. Through combination of agent behaviors and policy interventions, ABMs are also suitable for policy decision-making. Just like what UrbanSim model developed, we can build a urban system as Fig. 1.11 to combine the individual behavior with relevant urban development strategies.

It is supposed that different urban development strategies effects on agents' behaviors and finally result in macro results of urban boundary change, household mobility and location choices, location choices of tertiary sectors and relevant environmental impacts. If we take a certain policy set as model input, the agent behaviors affected by the policy set will be the core modeling contents. Following this logic, the environmental impacts will be the output of the modeling. If there are determined constraints for environmental impacts, it is possible to make a decision on the original inputs. For example to adjusting the policy set to make the water consumption of the whole system under the water supply capacity of case study area.

This book is designed on this logic, and from the next chapter the detailed information about how to build an ABM and how to utilize it to support targeted urban development strategy will be introduced. Basing on how an ABM is build that introduced in Sect. 3.3, Chap. 2 will introduce how to develop a environment for studying on urban development strategies. Chapter 3 is about the spatial process affected by regional spatial strategic plan. Basing on the urban sprawl, the household residential location choices affected by a certain residential policy will be introduced in Chap. 4 and the location choices of commercial shops and public facilities for elderly will be introduced in Chaps. 5 and 6. The Chap. 7 is about how to improve the split rate of public transportation in a city. And the last chapter a combined conceptual model of urban boundary change, household mobility and location choices, location choices of tertiary sectors and relevant environmental impacts will be introduced.

References

Batty M., Couclelis H., Eichen M., 1997. Urban systems as cellular automata, Environment and Planning B: Planning and Design. vol. 24(2), pp. 159–164.

Batty M., 2005. Cities and Complexity: Understanding Cities with Cellular Automata, Agent-Based Models, and Fractals. Cambridge, MA: The MIT Press.

Berger T., Schreinemachers P., Woelcke J. 2006. Multi-agent simulation for the targeting of development policies in less-favored areas. Agricultural Systems, 88(1), 28–43.

Bonabeau E. 2003. Predicting the Unpredictable. Harvard Business Review, 80, 109–116.

Chen L. Agent-based modeling in urban and architectural research: A brief literature review. 2012. Frontiers of Architectural Research, 1, 166–177.

Chinesta, F. I., 2021. Wealth Distribution and Inheritance: A Agent- Based simulation analysis.

Dawn C. P. and Filatova T. 2008. A conceptual design for a bilateral agent-based land market with heterogeneous economic agents. Computers, Environment and Urban Systems, 32, 454–463.

Dhou, K., Cruzen, C., 2021. An Innovative Employment of the NetLogo AIDS Model in Developing a New Chain Code for Compression. In International Conference on Computational Science (pp. 17–25). Springer, Cham.

Diappi L. and Bolchi P. 2006. Gentrification Waves in the Inner-City of Milan – A multi agent / cellular automata model based on Smith's Rent Gap theory, Van Leeuwen, J.P. and H.J.P.

Epstein, J., R. Pankajakshan, and R. A. Hammond, 2011. Combining Computational Fluid Dynamics and Agent-Based Modeling: A New Approach to Evacuation Planning. PLoS ONE, 6(5), e2013.

Ettema D. 2011. A multi-agent model of urban processes: Modelling relocation processes and price setting in housing markets. Computers, Environment and Urban Systems, 35, 1–11.

Klaassen, L. and Paelinck J. 1979. The feature of the large towns. Environment and Panning A, 11, 1095–1104.

Lee Q., Park B., Vlek S. J. P.L.G. 2010. Land Use Dynamic Simulator (LUDAS): A multi-agent system model for simulating spatio-temporal dynamics of coupled human–landscape system: Scenario-based application for impact assessment of land-use policies. Ecological Informatics, 5(3), 203–221.

Li X., Yeh A.G.O., 2000. Modeling sustainable urban development by the integration of constrained cellular automata and GIS, International Journal of Geographical Information Science. vol. 14, pp.131–152.

Jager W., Mosler. H. J. 2007. Simulating Human Behavior for Understanding and Managing Environmental Resource Use. Social Issues, 63(1), 97–116.

Ma Xinlin, Li Xijing, Kwan Mei-Po, Chai Yanwei, 2020. Who Could Not Avoid Exposure to High Levels of Residence-Based Pollution by Daily Mobility? Evidence of Air Pollution Exposure from the Perspective of the Neighborhood Effect Averaging Problem (NEAP), International journal of environmental research and public health, 17(4).

Ma Y., Shen Z.J., Kawakami M., 2013. Agent-Based Simulation of Residential Promoting Policy Effects on Downtown Revitalization, the Journal of Artificial Societies and Social Simulation, 16(2) 2.

Maes, P. 1995. Artificail Life Meets Entertainment: Life Like Autonomous Agents. Communications of the ACM, 38(11), 108–114.

Mitani T., Yamanaka H., 1996. An impact analysis of the development ordinance and the land area fund system on its urbanization—from the practice in Kanaiso District of Komatsujima City, Journal of the City Planning Institute of Japan (31), pp. 529-534 (in Japanese).

Moeckel R., Spiekermann K., Schürmann Carsten., Wegener M., 2003. Microsimulation of Land Use. International Journal of Urban Sciences, 7(1), 14–31.

Nakamura K. et al., 2004. A study on the analysis about the effect of the central area's regeneration measure considering the behavior of parking lots locating, Journal of the City Planning Institute of Japan, No. 39–3, pp.811–816 (in Japanese).

Noth, M., Borning, A., Waddell, P., 2003. An extensible, modular architecture for simulating urban development, transportation, and environmental impacts, Computers, Environment and Urban Systems, 27 (2), 181–203.

Q. B. Lee, S. J. Park, P.L.G. Vlek, 2010. Land Use Dynamic Simulator (LUDAS): A multi-agent system model for simulating spatio-temporal dynamics of coupled human–landscape system: Scenario-based application for impact assessment of land-use policies. Ecological Informatics, 5(3), 203–221.

Resnick, M., 1994. Turtles, Termites and Traffic Jams: Explorations in Massively Parallel Microworlds. Cambridge, MA: MIT Press.

Russell S. J., Norvig P. 1995. Artificial Intelligence: A Modern Approach. Prentice Hall.

Schelling, T. C., 1971. Dynamic Models of Segregation. Journal of Mathematical Sociology, 1(1), 143–186.

Shen Yue, Ta Na, Chai Yanwei, 2020. The Internet and the Space-Time Fixity and Flexibility of Daily Activities: A Case Study of Beijing, China, Cities, 97(C).

Shen Z., Kawakami M., 2008. Geo-simulation model using geographic automata for simulating land use patterns in urban partitions, Environment and planning B: planning and design. vol. 36, pp. 802–823.

Torrens, P.M., 2007. A geographic automata model of residential mobility, Environment and Planning B: Planning and Design, Vol. 34, No. 2, pp.200–222.

Torrens P.M., Benenson I., 2005. Geographic Automata Systems, International Journal of Geographical Information Science. vol. 19 (04), pp. 385–412.

Von Neumann, J., 1951. The general and logical theory of automata. In: Jeffress, L.A. (Ed.), Cerebral Mechanisms in Behavior – The Hixon Symposium. John Wiley and Sons, New York, pp. 1–31.

W. Jager, H. J. Mosler, 2007. Simulating Human Behavior for Understanding and Managing Environmental Resource Use, 63(1), 97–116.

White R. and Engelen G., 1997. Cellular automata as the basis of integrated dynamic regional modelling, Environment and Planning B: Planning and Design. vol. 24(2), pages 235–246.

Yu H., Russell A., Muhollamd J., Huang Z., 2018. Using Cell Phone Location to Assess Misclassification Errors in Air Pollution Exposure Estimation. Environmental Pollution, 233: 261–266.

Yu S. M., Ho K. H., 2006. Planned urban industrialization and its effect on urban industrial real estate valuation: The Singapore experience, Habitat International 30, pp. 509–539.

ZJ. Shen, Xiaobai A Yao, M. Kawakami, P. Chen, 2011. Simulating Spatial Market Share Patterns for Impacts Analysis of Large-Scale Shopping Centers on Downtown Revitalization, Environment and planning B: planning and design, 38(1): 142–162.

Chapter 2
The Environment for Accommodating Agents and Representing Urban Planning Conditions

2.1 Introduction

In last chapter we have already introduced the complexity of Urban system, such a complex urban system is an interconnected part that is generally defined as dynamic systems for exhibiting recognizable patterns of organization across spatial and temporal scales. (Parker et al. 2003). Researchers manage to understand it through many methods, modeling is one of them, which targets to represent the spatial process and future behavior through simulating the interactions between different parts of it. The interaction between different parts of an urban system also includes the policy interventions, such as planning policies. The new approach that aims to explore the impact of urban planning policies should be able to capture the phenomena that come into being with the performing of planning policy and can be able to recognize its common patterns and underlying mechanisms. The new ways of representing change in urban systems through rule-based decision process can be articulated through new ways of representing urban elements as individuals that are called "agents" and "cells" define their location (Batty 2005).

The ABMS are proved to be effective in decision-making support of policies. But many researches are theoretical, a few ABMs are developed to predict and evaluate the effects of a real world's policy scenarios and planning alternatives. Brown et al. (2004) presented an agent-based and analytical modeling to evaluate the effectiveness of a greenbelt. Their simulation results suggest how width and location of the greenbelt could help determine the effectiveness of greenbelts for slowing urban sprawl, but that these relationships are sensitive to the patterns of landscape aesthetic quality and assumptions about service center locations. The agent-based model developed by Kii and Doi (2005) is an example used to examine the effectiveness of policy measures aimed at achieving a compact city from the viewpoint of urban physical compactness, total trip length, energy consumption, and the social welfare of residents. Hanley and Hopkins (2007) reported a MAS model used to

Y. Ma, Z. Shen, *Strategic Spatial Planning Support System for Sustainable Development*, Advances in Geographic Information Science,
https://doi.org/10.1007/978-3-031-07543-8_2

assess the impact on single-family residential development patterns of plan for size, location, and timing of sewer line extensions, policies for extension timing, and of responses to these plans and policies by landowners and developers.

Such models are going to be more popular today due to the availability of individual-based data is improving. But besides the decision-making process is necessary in developing an ABM for evaluating and predicting policy scenarios, the reality of the urban space, agent behaviors and interactions are also necessary. As the basis of an ABM, the environment of an ABM should be the same with the real environment where individual behaves in reality. Thus, the environment is not just a sample cell that contains spatial attributes like location (coordinate) and areas, it would supposed to contain more complex spatial attributes for representing the real urban area. For example, if an ABM is developed to support decision-making on a residential policy, the environment should be able to represent the spatial characteristics of the urban area where the households are living in reality. If the logic for households' location choice is that households will evaluate the utilities of different location alternatives before make final decisions, the environment in the model hereby, should contain the special features of each land use parcel that households concerned for utility evaluation. Meanwhile, due to land use development in an urban area should follow determined spatial planning policies, such as urban master plan, urban land use zoning, ect. Thus, the environment in an ABM should also contains the spatial planning features too. In the next section the detailed information for how to create an environment for agent-based modelling of urban planning policy will be introduced.

2.2 Hypothetical Urban Space for Accommodating Urban Planning Conditions

2.2.1 An Environment for Representing Planning Urban Form

In practice it is possible to create different environment for an ABM. The environment is usually composed by cells in simulation model. Each cell has a determined projected coordinate (x, y) and an area. The coordinate (x, y) of a cell represents the location of that cell in an environment and each cell also has a spatial neighborhood feature. The spatial neighborhood feature of a cell means that taking the cell as a center of a region there will be a list of cells are close to it and recognized as the neighborhood of the cell. The traditional Von Neumann neighborhood is organized by a center cell and other 4 cells which are located on the east, south, west and north of the center cell (Fig. 2.1a). This neighborhood is extended to Moore neighborhood, which is composed by a center cell and other 8 cells (Fig. 2.1b) or 16 cells (Fig. 2.1c). The Fig. 2.1 shows the different characteristic of the neighborhood. The spatial neighborhood feature of cells in an environment supports us to conduct dynamic neighborhood analysis in a model and in the meantime, makes it possible

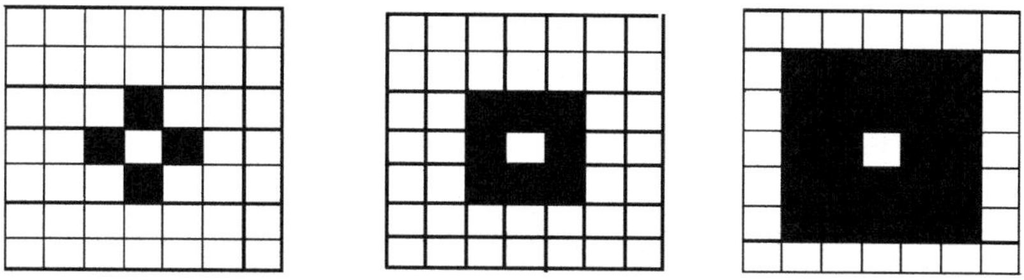

(a)Von Neumann neighborhood (b) Moore8 neighborhood (c) Moore16 neighborhood

Fig. 2.1 The neighborhood of cells in an environment

that to define the neighborhood of agents in the environment and simulate their behaviors on neighborhood level in and ABM.

Besides the spatial neighborhood feature of cells, if an ABM is designed to urban modeling, the environment could also be created to reflect different urban forms too. Actually in the practice of spatial strategic plan, planners generally need to determine the future urban form of a planning area first, the determined urban form will be a basis for its further spatial development control and land use zoning. The Fig. 2.2 shows the examples of created environment in Netlogo platform, which are the conceptualized urban forms of a planned urban area we frequently meet in planning practice. In this figure each cell represents a land use parcel of research area and cells could be connected by roads. The status of each cell could be stable on spatial-temporal and just for carrying agent's behaviors or dynamic for modeling the spatial process of urban development. The former could be a household location choice model, where agents make locations on cells according to the utility produced by each cell in a determined urban space. The latter could be a Cellular Automaton model for simulating urban growth/urban sprawl.

The urban form is greatly influenced by factors from two aspects, which are respectively the characteristics of natural resources, such as slope, hydrogeology, land resources, ect, and the historical social-economic development (including planning policies), planning policies for future urban spatial strategic plan. These factors are also parts of the important factors that influencing on land use suitability of urban development. Thus, it is possible to combine the conceptualized urban form (shown as Fig. 2.2) with the land use suitability evaluation to create an environment for representing planned urban area. In china, different land use types in an urban master plan must follow a range of proportions determined by City Planning Law. Figure 2.3 shows a combined system, in which the platform allows user to adjust the proportions of ten types of land uses in an urban planning area. The urban space of the system is represented by raster data and cell size is 50 * 50 m^2. By setting the land use proportions, the system will allocate land uses to cells basing one the suitability of each cell. This process will constraint by the planned urban form of that area, in this case the planning area follows the linear urban form. And last, the volume of carbon emission could be evaluated basing on the allocated land use of the planning area basing on Eq. (2.1).

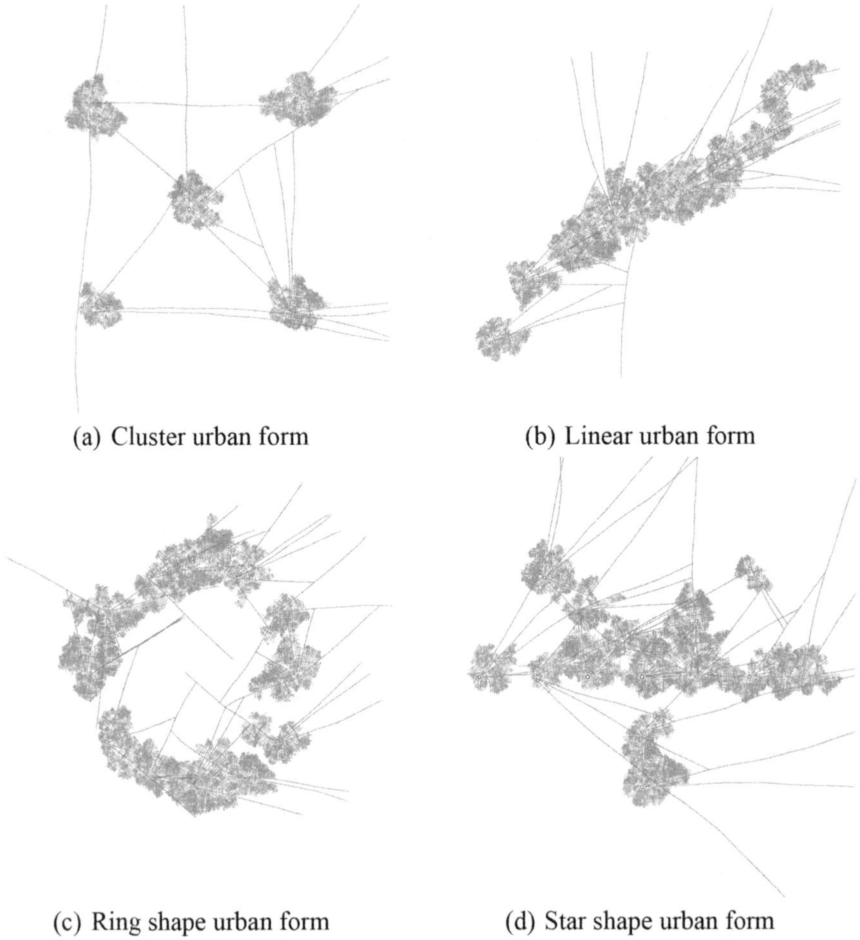

Fig. 2.2 An example of environment in ABM for representing different urban form

$$T = \sum S_i * p \quad (2.1)$$

Where T is the total volume of carbon emission of planning area, S_i stands for the area of land use i, and p is the carbon emission coefficient of land use i.

In this model, the total volume of carbon emission of planning area could be evaluated. It is supposed to be able to support government decision-making on low carbon spatial strategic plan. The evaluation of urban land use suitability and its coupling relationship with urban spatial change will be further introduced in next chapter.

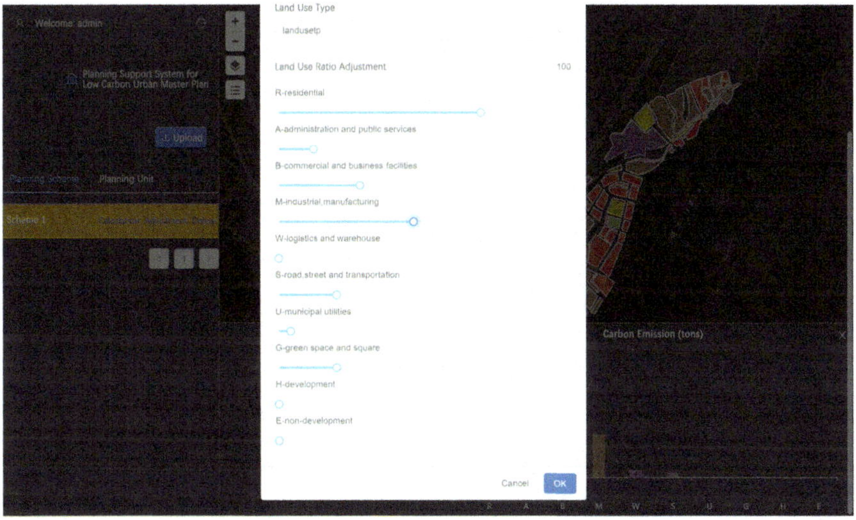

Fig. 2.3 Adjusting land use proportion to create an environment

2.2.2 An Environment for Representing the Urban Planning Condition

If an environment is going to be developed in an ABM for accommodating urban planning conditions, it means that the environment is adapting to a certain area and should follow its urban planning law. In this part a hypothetical urban space for accommodating urban planning conditions is created (Fig. 2.4).

To reflect the urban planning condition by the environment of an ABM, the Japanese City Planning Law is chosen. In Japan the city planning law for spatial control of cities are divided into two types. The first type is called Delineation City (DC) and the other one is Non-Delineation City. For the former, an Urbanization Control Line (UCL) is set to divide urban planning area into Urbanization Promoting Area (UPA) and Urbanization Control Area (UCA). UPA is the area where local government is willing to promote urbanization and land use zoning must be carried out. UCA is the area where urbanization must be constrained. Land use zone is further classified into three major groups: residential, commercial and industrial uses including 12 categories of land use zones.

For Non-Delineation City (NDC), urbanization promotion area and urbanization control area are not distinguished. But it also consists of two kinds of area, which are the Certain Uses Area and White Land area, respectively. The Certain Uses Area is classified according to how the land use parcels are already developed. While the White Land is about the land parcels where land use types have not been defined in current planning. Figure 2.5 shows the detailed planning conditions about the urban planning area in Japanese City Planning Law.

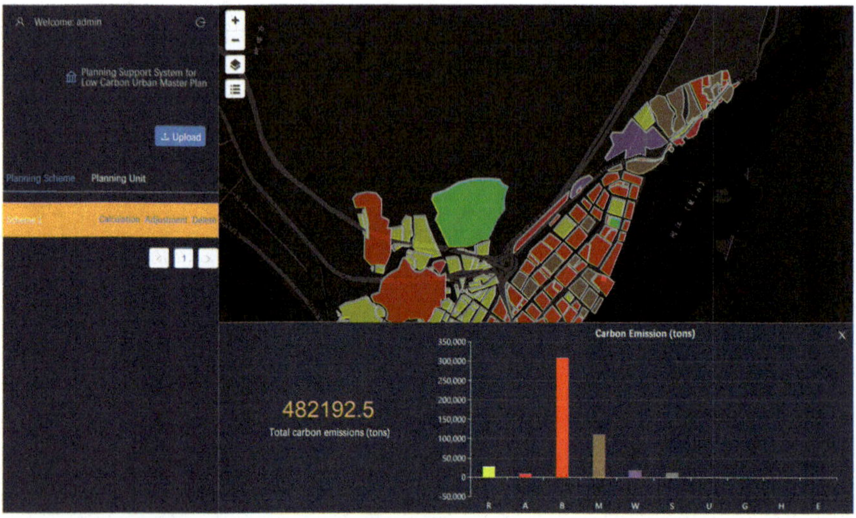

Fig. 2.4 Allocating of land use parcels in a planning area

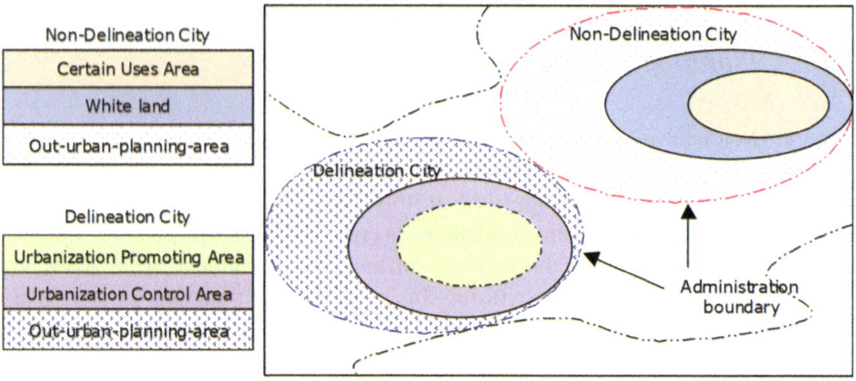

Fig. 2.5 Illustration of cities executing City Planning Law

Meanwhile, the defined environment assumes that the hypothetical urban space has following typical characteristics of delineation cities in Japan:

1. the urban space in urban planning area is divided into UPA and UCA;
2. there is a traditional central area (CA) located in the center of the city;
3. the land use zones within UPA are predefined according to Japanese city planning law;
4. the commercial area of the city is predefined that there are at least one large-scale shopping center (called B-shop) located outside of CA and several medium or small scale shops concentrating in CA (called S-shop).

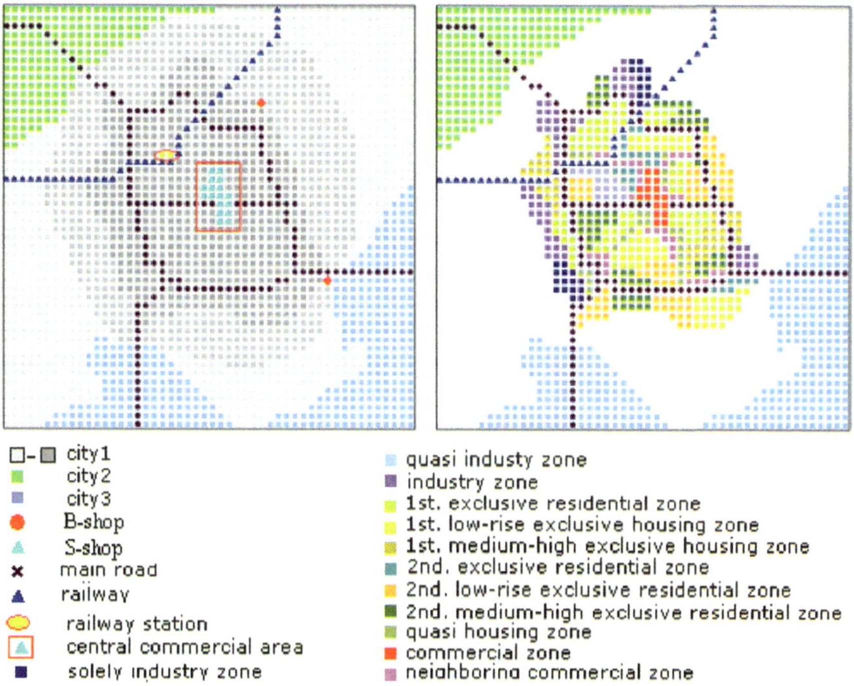

(a) Urban Space (b) Land use zone in central city

Fig. 2.6 The environment of simulation provided by planner agent

The virtual urban space is constituted by 2500 cells (50 × 50) where each cell measures 500 m × 500 m (Fig. 2.6a, b). This virtual urban space will be the environment of the ABMs going to be introduced through Chaps. 4, 5, and 6 in this book. The land use zones and relevant urban planning regulations will be the important constraints on agents' distributions, which would vary in different policy scenarios.

2.3 Distributing Agents on the Environment and its Validation

2.3.1 Factors Affecting Spatial Distribution

Due to the logic of ABMs development in this book is concentrating on the responses of households' behaviors affected by different spatial strategic plans, and hereafter, supporting the decision-making on relevant policies. The household behaviors are the core. Thus, the agent creation in the ABM should also follow the planning rule and regulations in case study area (Chen et al. 2007). In this part household agents will be distributed on the introduced environment according to the planning conditions in Japan. It is assumed that the number of household agent has been known and set by users in the ABM. Household agents are group into three income levels, which are respectively the Rich Households, the Middle-Income Households and the Poor Households. The sum of ratios of these three income households in a city equals to 1. It is assumed that there are three kinds of spatial pattern by income types, which are Well-Mixed, Inner-Higher and Outer-Higher, respectively. The Well-Mixed pattern represents an ideal household distribution pattern that different income households are mixed with each other. Different from the Well-Mixed pattern, the Inner-Higher and Outer-Higher pattern is defined according to the spatial aggregation of Rich households. The former means that the Rich household agents intend to live in the central city area, and the later vice versa. The ratios of different income-types could be adjusted by users and select the spatial pattern of households' distribution for the further simulation, such as simulation of shop location choices.

2.3.2 *Land Use Zoning Constraints*

The land use zoning type of a cell in CA and UPA is given to values randomly in the range of 1–12 representing 12 types of land use in Japan (Table 2.1). Before the simulation started, the value of cells in UCA is given to 0. It means the land use zoning just happen in CA and UPA areas and once the zoning has be done randomly it will not change during the simulations.

There will be a variable namely HD to constraint the household distribution on different land use zones. The HD is one of the attributes of the cell, which stands for the household density in reality. Due to the cell size are all the same in the

Table 2.1 The zoning constrains related variables for the UPA

No.	Land use zoning	The central area HSR (%)	The central area HC (%)	Out-The central area HSR (%)	Out-The central area HC (%)
1	1st low-rise exclusive residential district	100	2	100	1
2	2nd low-rise exclusive residential district	100	2	100	1
3	1st mid-high-rise exclusive residential district	100	3	100	2
4	2nd mid-high-rise exclusive residential district	100	3	100	2
5	1st residential district	90	4	90	2
6	2nd residential district	90	4	90	2
7	Quasi-residential district	80	4	80	2
8	Neighborhood commercial district	70	4	70	2
9	Commercial district	50	10	50	5
10	Quasi-industrial district	70	4	70	2
11	Industrial district	30	4	30	2
12	Exclusive industrial district	0	0	0	0

environment of an ABM, so that the HD will be set as a number to represent the maximum number of household agents on a cell. The number of household agents (Hnum) living in a cell cannot exceed the HD of the cell. The value of HD for each cell in urban planning area is set from the maximum floor area ratio (FAR) and the housing use ratio (HSR). These two variables are varied from different land use zones so that the HD of each cell is also different and related to the land use zones where the cell belongs to.

The detailed value of these two variables in each kind of zoning types in real world is referred to the existing study by Hiroshi Kidani and Mitsuhiko Kawakami in 1996. In order to simulate the general spatial distribution pattern of population density in real cities, it is assumed that the value of HD of cells in CA is higher than other cells belonging to UPA (as shown as Table 2.1). In planning practice although the land use zones has not been conducted in UCA, it does not mean there are no residents. Conversely, there are often residents live in the UCA of a local city. But it is difficult to assess the probability of a place in the UCA to be developed. Regarding to this situation the HD of a cell in UCA is set either 1 or 0 with equal probability in the virtual urban space.

2.3.3 Housing Suitability of Each Cell

In reality, the distribution of households in an urban space is the results of household residential location choices. Thus, the created virtual urban space (environment) should be able to reflect the suitability of each cell for household agents' distribution. The environment developed in this chapter also concludes this spatial attribute namely housing suitability. It represents the physical environment conditions of a cell that affecting household distribution. To evaluate housing suitability is normally a complex process because many factors will affect housing suitability, such as slope, transportation, public facilities, land use zones and developed land uses ect. While in this chapter the housing suitability for the virtual urban space is given randomly before starting the simulation. There are four ranks of the housing suitability (S-level): High, Middle-high, Middle and Low, respectively numbered as 4, 3, 2, 1 for cell attribute values. The detailed housing suitability evaluation on residential location choices will be introduced in Chap. 4.

2.4 How HDF Works in an ABM

To locate household agents in the determined environment, a simulation flow is needed. It is named Household Distribution Flow. The HDF works at the preliminary initialization step of an ABM. The household distributions emerging from it will work as a precondition for further simulation of agent behaviors, such as shop-choice behaviors or residential location choice behaviors. To carry out household

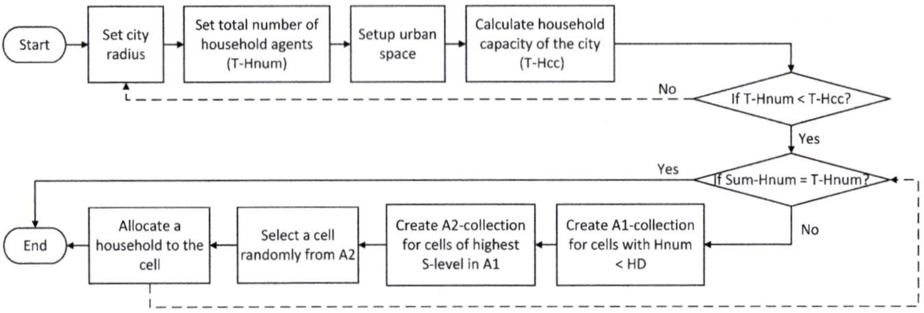

Fig. 2.7 The flow chart of household distribution on designed environment

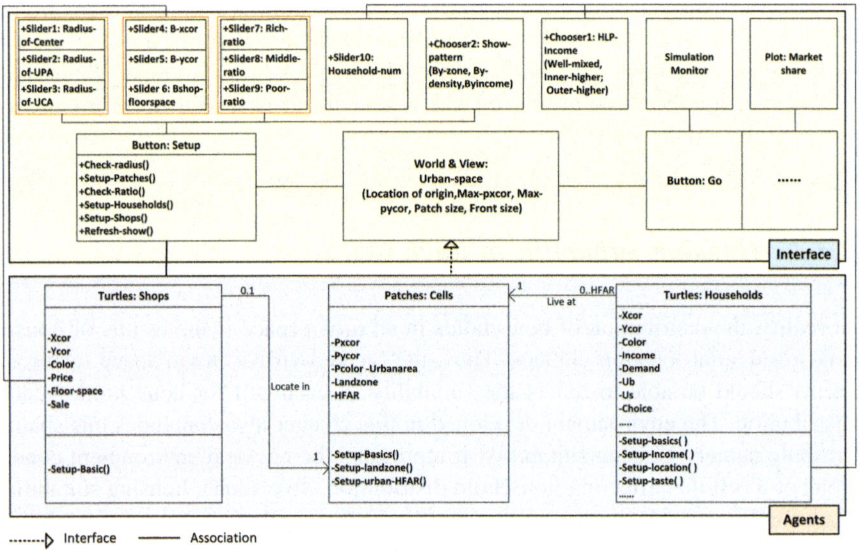

Fig. 2.8 The class diagram of HDF in Shopsim

distribution model, users should set some global variables' values, which include radiuses of the three sub-areas, the total household number (THnum), and spatial location pattern of different income households. The Fig. 2.7 shows the simulation flow of household distributing household agents on designed environment in an ABM, which is also the virtual urban space introduced in Sect. 2.3. The Fig. 2.8 demonstrates the relationship between the environment and household distribution in an ABM.

2.5 Validation of HDF for Urban Planning Conditions

2.5.1 *The Variogram Model Using for Validation of HDF*

The most important characteristic of HDF is that given the total number of households, instead of placing household randomly, the HDF enable to locate household one by one on the environment where urban planning conditions are combined. The conditions include the sub-area radiuses which decide the size of urban planning area, the land use zoning type and the housing suitability of cells. Due to the sub-area radiuses can be planned so that the sub-area each cell belongs to can be controlled outside the model. Meanwhile, the land use zones and housing suitability are regarded as known conditions to a virtual city. Furthermore, the HDF can generate different household distribution patterns no matter whether the city radiuses are same or not. There is a need to validate the household distribution model to confirm whether it can produce household distributions with certain observable regularity due to the change of urban planning conditions that can be controlled outside the model, here the sub-area conditions in a scientifically meaningful manner. Here we propose a use of variogram as a method of HDF validation.

The variogram is a fundamental geo-statistical technique to depict the spatial correlation structure in the sampled spatial data (Houlding 2000). The variogram analysis consists of the empirical variogram calculated from the data and the variogram model fitted to the data. In the analysis, the empirical variogram is usually calculated as semi-variograms and plotted as a 2-dimensional graph. The variogram models describe spatial variability using the parameters 'range', 'partial sill', 'nugget' and 'sill' to fit empirical variogram values derived from the data. The 'sill' is the sum of 'partial sill' and the 'nugget'. So, analyzing the parameter features of 'range' and 'sill' can provide an insight into the spatial characteristic of the data under studying. In this paper, the Spherical variogram model is used to validate household distribution flow, which is illustrated as Fig. 2.9.

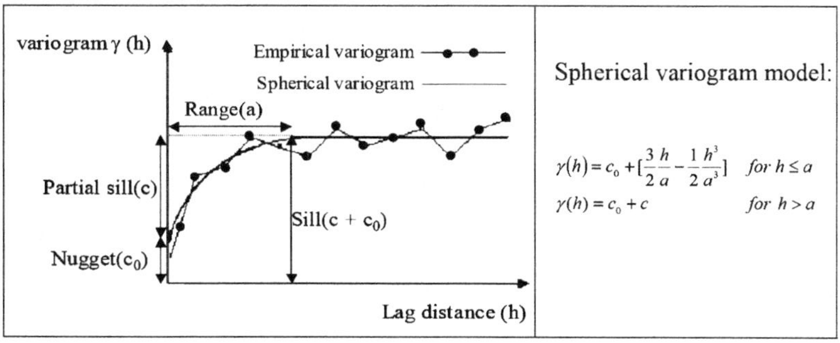

Fig. 2.9 The illustration for variogram

2.5.2 *Sub-area Radiuses Using as Validation Parameters*

For validating the HDF, an ideal monocentric virtual city is used. Its urban planning area is divided into three sub-areas, which are the Central Area (CA), the Urbanization Promotion Area (UPA) and the Urbanization Control Area (UCA), respectively (as shown as Fig. 2.4). The household can only be allocated to sub-areas of CA and UPA while considering urban planning conditions that are sub-area radiuses, the land use zoning types and the housing suitability. In this chapter, sub-area radiuses of the virtual urban space are taken as the test parameter and can be changed to generate different household distributions. A random method is used to set the land use zoning conditions and housing suitability before starting simulations, and they will keep as a constant during simulations.

The 2-dimensional latticed space is composed of 2601 square cells (51 × 51) where the side-length of one cell is one unit's distance. The size of each sub-area is decided by three sub-area radiuses: the radius of CA (R1), the radius of UPA (R2) and radius of UCA (R3) (as shown as Fig. 2.10). If R2 is the same as R1, the cells within the radius of the UPA will all have the same features as the cells in the CA, in other words, no cell belongs to UPA anymore. If R2 is set equal to R3, then the sub-area UCA will disappear and cells fall between R1 and R2 (R3) will be all looked as cells of the UPA. The size of urban planning area is decided by the R3. For convenience, the R3 is set as a constant value of 25 that means the urban planning area consists of 1961 cells.

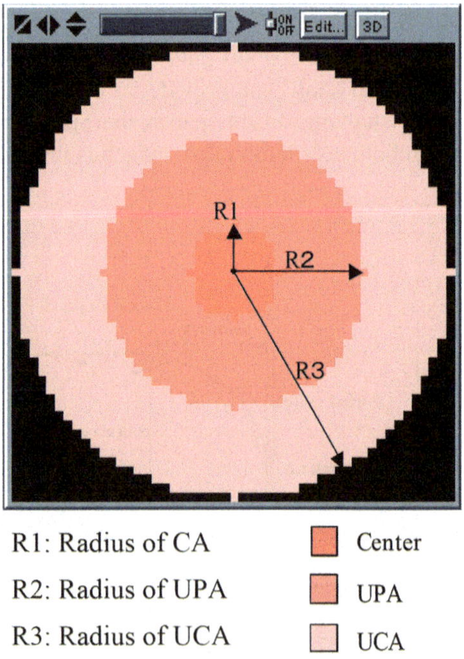

Fig. 2.10 The illustration for sub-area radiuses

For validation, five sets of values (T1, T2, T3, T4, T5) are given to sub-area radiuses as listed in Table 2.2. The total household number is set as 1500 and keep the same within all simulations. For convenience, the household spatial distribution results that variograms generated with these five sets of parameter values are also named T1, T2, T3, T4 and T5 respectively.

As mentioned above, the city radius will affect the scale of each sub-area, measured by the number of cells (see Table 2.3). Comparing T1 and T2, R2 is unchanged, and when increasing the value of CA radius, the CA area will become larger and the UPA area will become smaller correspondingly. However, comparing T2 and T3, it can be seen that when R1 is kept stable, raising R2 will lead more cells in the UPA change to the UCA. T4 is a special case in which there is no cell belonging to sub-area of UPA, and T5 is another one in which no cell belongs to sub-area of UCA. Due to household agents can only be allocated within the radius of the UPA and the household density of cells in different areas are different even with the same land use zones, so the household distributions are supposed to be also affected.

The household distributions generated from the HDF using these five sets of sub-area radiuses are showed in Fig. 2.11. It can be seen that the household density of the sub-areas is different from T1 to T5, but it is hard to know whether differences between distributions are caused by the different values of sub-area radiuses instead of the random variables of the model. It is also difficult to measure general differences quantitatively even if these differences are caused by the change of city-radiuses. In the next section, the variogram method is described to analyze the characteristic of household distributions and quantify the influence of the sub-area radiuses on the output of the HDF.

2.5.3 *Using the Variogram for Validation*

To validate the HDF, with each set of parameters, the household distribution model is performed 30 times in total, and the distribution data of 30 times are recorded into a data file. Each row of the data file recorded the household distribution status of

Table 2.2 Sub-area radiuses for household distribution model

Radius	T1	T2	T3	T4	T5
R1	5	10	10	15	5
R2	15	15	20	15	25
R3	25	25	25	25	25

Table 2.3 Number of cells in sub-areas of different radiuses

Sub-area	T1	T2	T3	T4	T5
Center	81	317	317	709	81
UPA	628	392	940	0	1880
UCA	1252	1252	704	1252	0
Sum	1961	1961	1961	1961	1961

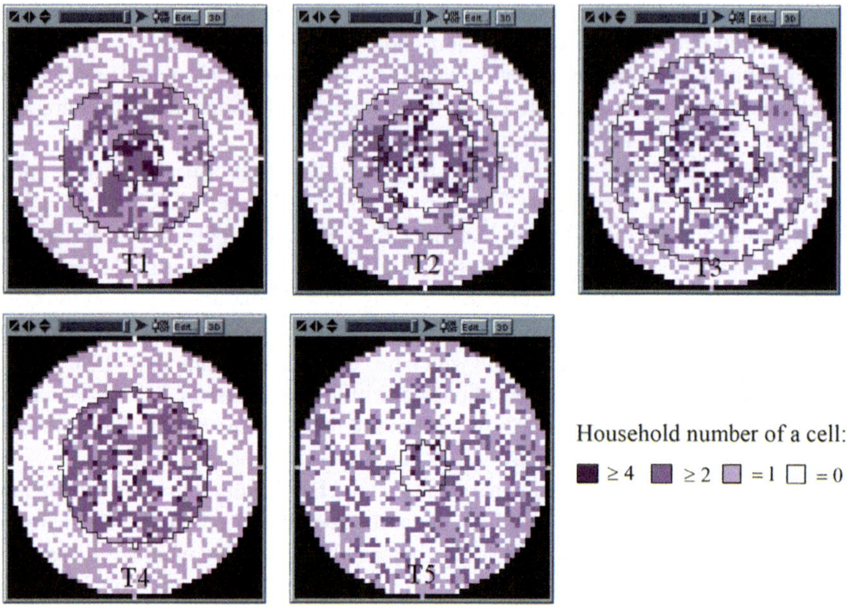

Fig. 2.11 The spatial distribution of household agents

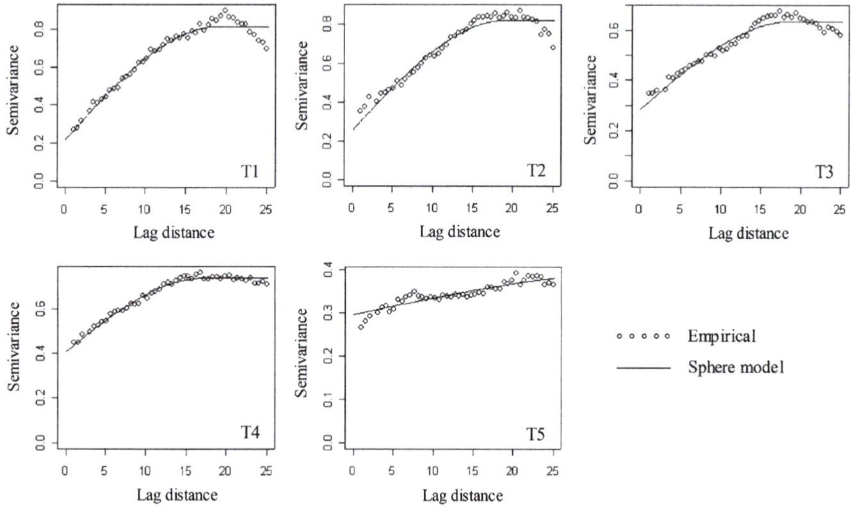

Fig. 2.12 Sphere variogram curves fitted to empirical variograms

each times including the coordinates, sub-area, the land use zoning type, HD of each cell, and household number on each cell. These five files of household distribution data (T1, T2, T3, T4, T5) are read into software R for variogram analysis. The variogram functions and commands available in the geoR package of R is used for this work. Corresponding to each household distribution data, 30 empirical variograms are computed and plotted.

Figure 2.12 shows the typical empirical variogram for each set of data. And 30 sets of parameters (sill, range) for the fitted spherical variogram models are estimated. The distribution of estimated sill and range for five data sets are summarized in a box plot, displayed in Fig. 2.13. The line in one box indicates the median value of sills or ranges. The means of sills and ranges are showed in the Fig. 2.14. Both from the Figs. 2.13 and 2.14, it can be seen that the values of sill and range are different in household distribution data. Each pair of sill and range will decide one

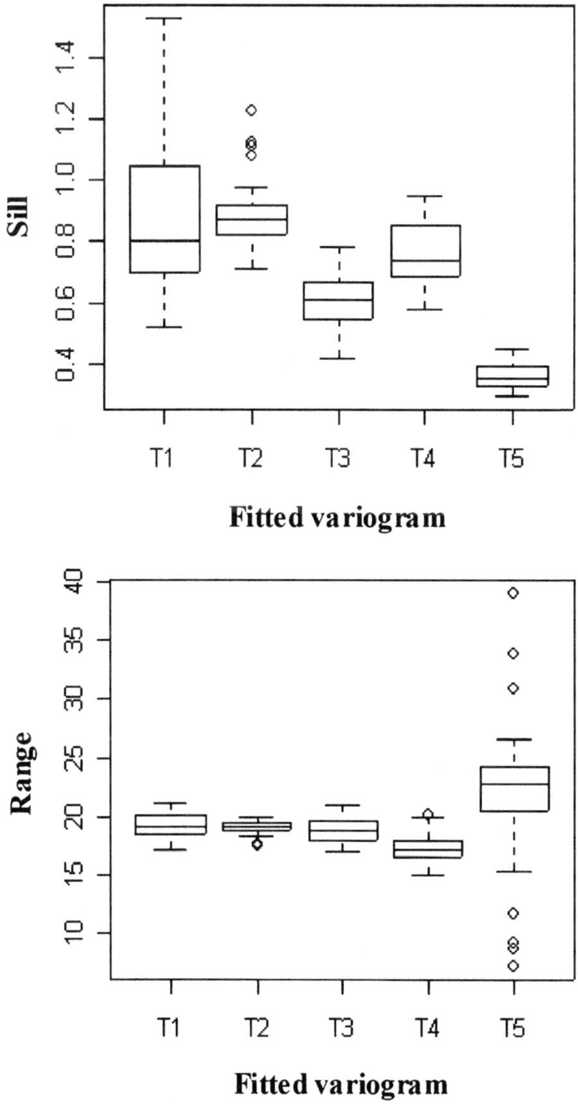

Fig. 2.13 The box plot of sills and ranges for fitted variograms

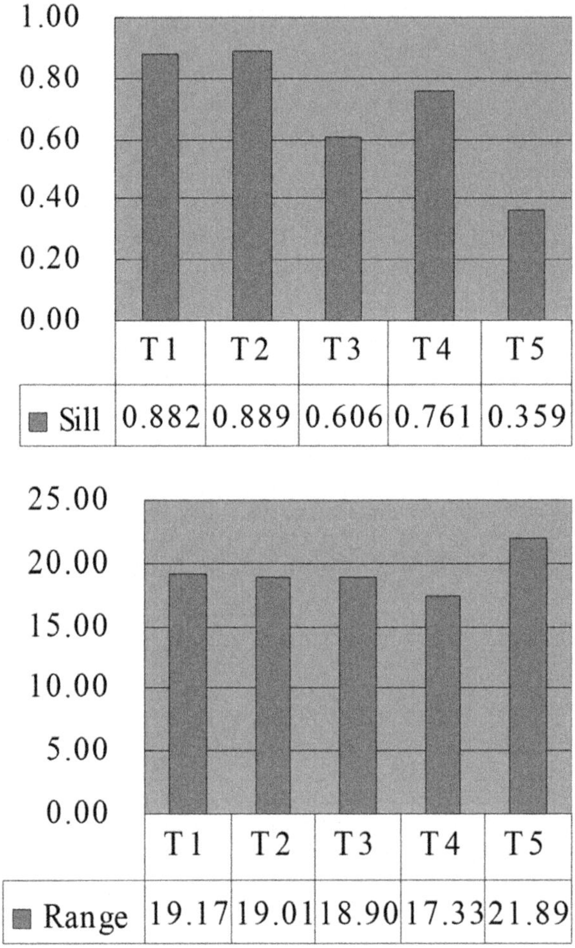

Fig. 2.14 The means of sill and range for fitted variograms

special fitted variogram and reflects spatial characteristic of household distribution. It indicates that the household distribution is sensitive to the sub-area radiuses changes. To investigate the general difference between five groups of data, a Scheffe's multiple comparison is done. The mean difference is set significant at the 0.05 (the Alpha Level). The result is showed in Table 2.4, which statistically proves that there are general differences between groups.

There are obvious differences in both or either means of sill and range between two groups except for T1 and T2. These may indicate that keeping the UPA radius stable and raising CA radiuses a little will not cause much difference in household distributions. But if further observing the mean differences of other groups, it could be found that there are significant differences between other groups' sill mean and the significant range mean differences are only exit in groups concerning T5. This

Table 2.4 Scheffe's multiple comparison for mean difference

		Sill		Range	
Ti	Tj	Mean Difference (Ti-Tj)	Sig.	Mean Difference (Ti-Tj)	Sig.
T1	T2	−0.0074	¬	0.1591	¬
	T3	0.2762	*	0.2654	¬
	T4	0.1207	*	1.8332	¬
	T5	0.5233	*	−2.7269	*
T2	T1	0.0074	¬	−0.1591	¬
	T3	0.2863	*	0.1063	¬
	T4	0.1280	*	1.6741	¬
	T5	0.5307	*	−2.886	*
T3	T1	−0.2762	*	−0.2654	¬
	T2	−0.2863	*	−0.1063	¬
	T4	−0.1556	*	1.5678	¬
	T5	0.2471	*	−2.9923	*
T4	T1	−0.1207	*	−1.8332	¬
	T2	−0.1280	*	−1.6741	¬
	T3	0.1556	*	−1.5678	¬
	T5	0.4027	*	−4.5601	*
T5	T1	−0.5233	*	2.7269	*
	T2	−0.5307	*	2.8860	*
	T3	−0.2471	*	2.9923	*
	T4	−0.4027	*	4.5601	*

The mean difference is significant at the 0.05 level; asterisk "*" donates to the difference is significant; mini-line "¬" means no significant difference

result indicates that the spatial variation level of household distributions generated from different set of city radius are general different. The range of the spatial correlation are going to be the same except for the extreme situation like T5 in which the UPA radius is set equal to the UCA radius and UCA is overlapped by UPA.

Basing on the above discussion, it can be proposed that household distribution is highly affected by the size of urban planning area. The value of sub-area radiuses also impacts on the output of the HDF too.

2.6 Discussions

In order to develop an ABM closely to reality, the environment in the model should keep the same characteristics with real. If the model is developed for decision-making of spatial strategic plans, the environment should contain the special characteristics of the planning area and reflects relevant planning conditions. Meanwhile, the agent's allocation in the ABM also needs to follow the planning conditions. In this Chapter a household distribution flow is designed to allocate household agents while considering urban planning conditions aiming enable the final simulation results of come closer to reality. The household distribution flow is validated by a geostatistical method called variogram.

The variogram approach to validate the household distribution model is an exploratory method of validation analysis for the MAS model. The variogram-validation of household distribution model was accomplished in two steps: first to produce empirical variograms for each household distribution data and then to estimate and compare the sill and range parameters for fitted variograms of household distributions. The results of the validation show that the sub-area radiuses have significant impact on the output of household distribution model and household distribution model can generate different household distributions with certain regularities for further simulation.

Obviously, there are still many problems left for further research. First, how to generate proper settings of the land use zoning type and land use suitability for virtual urban space is still unsolved, such as housing suitability for household distribution. Furthermore, how these settings will affect the household distributions still needs to be explored.

References

Batty, M., 2005. Agents, cells, and cities: New representational models for simulating multiscale urban dynamics, Environment and Planning A 37 (8), pp. 1373–1394.

Chen Ping, Shen Zhenjiang, Mitsuhiko Kawakami, 2007. Validation of household distribution model employed in multi-agent simulation using variogram—From a viewpoint of urban planning conditions, The 30th symposium on computer technology of information, systems and applications, Architectural Institute of Japan, pp.31–36.

Daniel G. Brown, Page, S. E., et.al, 2004. Agent-based and analytical modeling to evaluate the effectiveness of greenbelts, Environmental Modelling & Software 19, pp.1097–1109.

Hiroshi Kidani, Mitsuhiko Kawakami, 1996. Actual condition and planning theme of building activity in urbanization control area, Journal of the City Planning Institute of Japan, No.31, 583–588.

Houlding S. 2000. Practical Geostatistics: Modeling and Spatial Analysis 1. Springer.

Kii M, Doi K, 2005. Multiagent land-use and transport model for the policy evaluation of a compact city, Environment and Planning B: Planning and Design 32(4) pp. 485–504.

Parker, D.C., Manson, S.M., Janssen, M.A., Hoffmann, M.J., Deadman, P., 2003. Multi-agent systems for the simulation of land-use and land-cover change: A review, Annals of the Association of American Geographers, 93 (2), pp. 314–337.

Paul F Hanley, Lewis D Hopkins, 2007. Do sewer extension plans affect urban development? A multiagent simulation, Environment and Planning, 34(1), pp. 6–27.

Chapter 3
Simulation of Urban Growth and Household Aggregation for Planning Support of Local Spatial Strategic Plan

3.1 Introduction

In last Chapter we have illustrated how to create an environment during urban modeling process. This environment contains planning conditions not only about spatial planning and control but also population scale on each cell. During application of an ABM built on this environment, the environment could be stable or dynamic. If the planning policy will be launched in a city where the urban growth is not taking place frequently, it is possible to ignore the urban spatial dynamic during the policy application and keep the environment stable in the ABM. But if the planning policy would lead to urban spatial dynamic and correspondingly effects on agent behaviors, the changes of urban boundary and land use need to be considered in the model.

In this case we will introduce how to simulate the evolution process of the environment in ABMs during urban modelling. We will take local spatial strategic plan as an example to introduce how the urban spatial dynamic is simulated for the planning support of the plan. The developed land use cells in simulation will be seemed as the environment for accommodating agents. The agents in such models can be seemed as persons or households, of course the latter could be composed by a single person or several persons. In the model the environment for accommodating agents is dynamically growing and agents will be distributed on the developed cells basing on planning conditions, such as the constraint of population density determined by local urban master plan. But in this chapter the agents are only distributed on the simulated environment (developed urban space), the mobility of the agents will not be simulated. The detailed methods for simulating agents' behaviors in an urban space will be introduced in the following chapters.

In last decades, population migration in urban area is taking place at an unprecedented pace around the world and especially in developing countries, which usually leads to the urban growth and urban spatial dynamic (National Bureau of Statistic 1996–2005). Many researchers managed to reveal the relationship between

Y. Ma, Z. Shen, *Strategic Spatial Planning Support System for Sustainable Development*, Advances in Geographic Information Science,
https://doi.org/10.1007/978-3-031-07543-8_3

urban land use/cover change and household behaviors (Marwa et al. 2015; Fontaine and Rounsevell 2009; Han et al. 2009; Brown et al. 2006). These researches help us to understand the interactive process between land use/cover change and human behaviors, to some extent, they explain the process of urban growth. Thus, some questions are cleared that how urban growth taking place over time and what constraints effect on this process (Wu 1998). These answers make it possible that using models to forecast the process of urban growth or urban spatial dynamic (Barredo et al. 2003; Shen and Kawakami 2008).

There are various methods for representing urban growth or urban spatial dynamic, such as Land Use / Cover Change (LUCC) models, land use and transportation models, Cellular Automata (CA) models, agent-based models, et al. Although LUCC models or land use transportation models are good at reflecting the process of land use change and recognizing driving forces, they always lack in reflecting the planning constraints on urban growth and the dynamics of urban growth (Batty et al. 1997, 1999; Clarke and Gaydos 1998). CA approach has been proved that it has advantages and sufficient capability in simulating the dynamics of urban space (White and Engelen 1997; Wu 1998; Li and Yeh 2000; Fang et al. 2005). While, urban growth is a complex process that is influenced by many driving forces. The pure CA model is inadequate to taking political, economic, cultural, and natural driving forces into account (Ward et al. 2000).

In order to reflect the effects of above driving forces, the Constraint CA (CCA) model was developed for simulating the complex spatial process of urban growth (Li and Yeh 2000). The constrained CA model has been developed by combining the driving forces with the transition rules between automatons. The transition rules in a CCA model, helps automatons transfer information between neighbors by diffusion (Torrens and Benenson 2005; Torrens 2007). When utilized it in urban modeling, automatons could be seemed as a cell of spatial unit or a land use parcel. Therefore, urban growth simulation by using CA or CCA models look like assignment of the cells by the predefined neighborhood rules. In this way, it is impossible to reflect individual decision-making processes by using CA approach. Therefore, CA-based models have strong ability to simulate the spatial processes of urban growth affected by special driving forces, such as planning policies, local natural resources, ect. Basing on the simulation results of spatial process of urban growth, Agent-Based Model can be further developed and coupled to simulate individual decisions during urban growth.

Usually there are many pre-defined parameters in the CA model to represent the transition rule between automatons. The CCA model comparing with pure CA, driving forces on land use change are added in the transition rule. But the transition rule, or we call it the information diffusion between automaton neighbors is still pre-defined both in CCA or CA models. To evaluating the effects of each driving forces are usually based on logistic regression (Poelmans and Rompaey 2010; Munshi et al. 2014). However, logistic regression is hard to reflect the relationships between variables or driving forces and their complexity in land use change process. Although during model calibration we will validate developed model with real dataset, the validation is taking place between simulation results and real dataset, instead of the predefined transition rules and real rules. It may be resulted from no enough examples support us to recognize or automatically calculate the potential of status changes between automatons. While the development and application of Artificial Neural Network (ANN) Algorithm could make up this limit and support CA parameter calibration (Cao et al. 2014, 2016; Li et al. 2013). Consequently, the simulation of urban growth by using CA model is gradually combined with multiple driving forces (Ma et al. 2010) ANN algorithm (Saa and Liu 2022) or traditional Land Use/Cover Change (LUCC) model such as Markov chain (Han et al. 2015).

But the pure urban growth simulation can not be simply equal to planning support of special policies. The former could just concentrate on reveal the phenomenon of urban growth process and its future trend but weak the policy effects. While the latter should concentrate on reflect the effects of policy intervention on future trend, and therefore, the variables for transition rule should be able to represent relevant planning policies. In the following sections we will introduce two examples on how to utilize combined CA model to support local planning practice in China.

3.2 CCA Model for Representing Multi-driving Forces in Urban Spatial Dynamic Simulation

3.2.1 Combined System for Decision-Making Support of Local Spatial Strategic Plan

Before the territory spatial planning has been launched in China, the spatial planning works are divided into land use planning and urban planning, two parts. The land use planning responsible for the planning and management of territorial land resources both in urban and rural areas. The urban planning only concentrates on the land use and zonings in the urban planning areas. The two kinds of planning work are drafted by different government departments. The spatial strategic plan we mentioned here actually is the upper-level plan of urban master plan, which targets to determine the urban function, spatial structure, developing scale (population and developed land use) during future urban development.

The purpose of the spatial strategic plan makes it relate to economy and social development, land resources and urban planning. While these items are further related to different government departments. For example, urban function is determined by the industry of a city while planned by a government department namely committee of development and reform. The scale of developed land use needs to be coordinated to the planning policy that issued by original Land and Resources Bureau and urban and rural planning bureau (these two departments in most Chinese cities are combined into a new department, namely natural resources bureau). Thus, for decision-making of such a top-down planning practice, the planning policies need to be input as predefined parameters in simulation model. Furthermore, the model should be able to combine the planning policies issued by independent government departments.

Thus, in this work we utilized the CCA model that built by Wu, Li and Yeh to simulate the urban spatial dynamic process that effected by local spatial strategic plan. In the model, the constraints for urban development are sourced from 4 aspects, which are spatial constraint, macro-social economic constraint, institutional constraint, and neighborhood constraint (Long et al. 2008, 2009), respectively. These constraints are related to different planning policies. For example, the institutional constraint can be assumed as the urban land use zoning issued by master plan, the macro-social economic constraint is defined from the local national economy and social development plan. Additionally, to help land use cells to diffuse developing information between automaton neighbors, Moor4 neighborhood is utilized and each cell in the model is defined to have four neighbors. The logic of the simulation is showed by Fig. 3.1 and the detailed transition rule of cells is showed by Eq. (3.1).

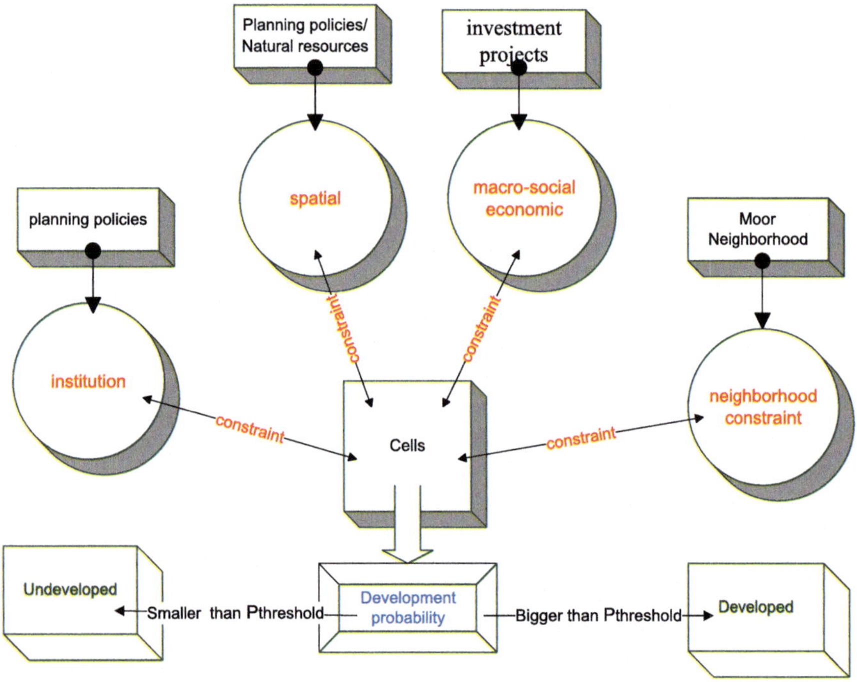

Fig. 3.1 the logic of urban spatial dynamic simulation by Constrained CA

$$
\begin{aligned}
& s_{ij}^{\,t} = x_0 + \sum_{n-1}^{k=1} x_k \times a_k + x_n \times a_n^{\,t} = s_0 + x_n \times a_n^{\,t} \\
& p_g^{\,t} = \frac{1}{1+e^{-s_{ij}^{\,t}}} \\
& p_{ij}^{\,t} = \exp\left[\alpha\left(\frac{p_g^{\,t}}{p_{g\,\max}^{\,t}} - 1\right)\right] \\
& \text{if } p_{ij}^{\,t} \geq p_{threshold}\left(p_{ij}^{\,t}, Development\text{stage}^{t+1}\right) S^{t+1}(ij) = \text{Developed} \\
& \qquad \text{otherwise } S^{t+i}(ij) = \text{UnDeveloped}
\end{aligned}
\tag{3.1}
$$

In the equation, a variable s_{ij}^{t} is utilized to represent the land use suitability of cell (i, j) at time t. x_0 is a constant and s_0 stands for the land use suitability produced by a cell's original conditions on natural resources and environment during each simulation step. $S^{t+i}(ij)$ is the status of cell (i, j)'s land use suitability at time t + 1.x_k is a parameter set reflecting institutional policies besides neighborhood rule. x_n is a parameter reflecting the land use suitability influenced by cell's neighborhood, a_n is spatial features of neighborhood, a_k are spatial features besides neighborhood, and p_g^t is global probability of cells transfer to developed land use at time t. $p_{g\max}^t$ is the max global probability during each simulation loop. α is diffusion coefficient which isintherangeof(1–10).p_{ij}^{t}isthefinalprobabilityang $p_{threshold}\left(p_{ij}^{t}, Development\text{stage}^{t+1}\right)$ is the neighborhood of cell number and developing speed control, which can change according to p_{ij}^{t} and $Dvelopment\text{stage}^{t+1}$ in order to make the developing speed in $Dvelopment\text{stage}^{t+1}$ is the same as settings.

The transition rule in the model is defined that if the final probability of each cell is bigger than the threshold, which is named as $p_{threshold}(p_{ij}^{t}, Development\text{stage}^{t+1})$ in the equation, the status of the cell will change to developed cell, vise verse.

After the undeveloped cells change to developed cells, the urban spatial dynamic process is simulated. While urban spatial dynamic will not only result to the spatial change of urban areas but also population aggregation. In this work, population that aggregate to developed urban areas are seemed as household agents. we assumed that when the agents aggerate to urban area, they will follow the regulation of urban spatial planning, such as the land use zoning. Thus, we further forecast the possible population scale effected by local spatial strategic plan. However, we have not simulated population mobility process and just generated household agents on cells based on planned population density of each land use zone. The logic for generating agents is showed by Eq. (3.2):

$$
\begin{aligned}
Nh = \Sigma_{x=r,c} Nx = & \left(\Sigma_{n=1} Drn * \varepsilon rn\right) \times Scell \times \Sigma NumCellr \\
& + \Sigma_{n=1}^{2}\left(Dcn * \varepsilon cn\right) \times Scell \times \Sigma NumCellc
\end{aligned}
\tag{3.2}
$$

in which, N_r, N_c and N_h, stand for the agent numbers in residential zone, commercial zone, and all developed cells, respectively. D_{rn} and D_{cn} stand for the planned population density of residential and commercial zones; ε is a random value with means of 1 and standard deviation of 0.5 in order to generate population density of each land use zone; S_{cell} means the area of each land use cell and $\sum NumCell$ is the total cell numbers of developed land use for each land use type.

3.2.2 *The Framework of the Combined System*

As shown in Fig. 3.2, the framework is formed by three parts, which are respectively data importing, data processing and data output. In the first part, the data of spatial patterns (land use zones), geo-surface (slope of research area) and initial development areas (locations of economic investments) will be imported from GIS database to the simulation. Among these data, the spatial pattern is referred to urban master plan. The Geo-surface is based on the slope data of research area and will be utilized to evaluate the land use suitability of each cell. In this work the slope data is sourced from 30-meter resolution remote-sensing data. It is supposed that other dataset could be supported by different government department. In this case, the planning work of different government departments could be combined in one platform for decision-making support. Accompanying with the reform of territory spatial planning in China, some of the planning practice showed in Fig. 3.2 has already changed but before the reform, such kind combined system is very necessary for decision-making of the strategic plans that related to planning policies issued by more than one government department.

The system follows a simulation flow as shown as Fig. 3.3. Firstly, initial parameters and data sources need to be setup. These parameters are used to reflect planned land use zones in research area. In the meantime, data sources relate to cell's spatial attributes such as land price, slope and initial development potentials will also be set

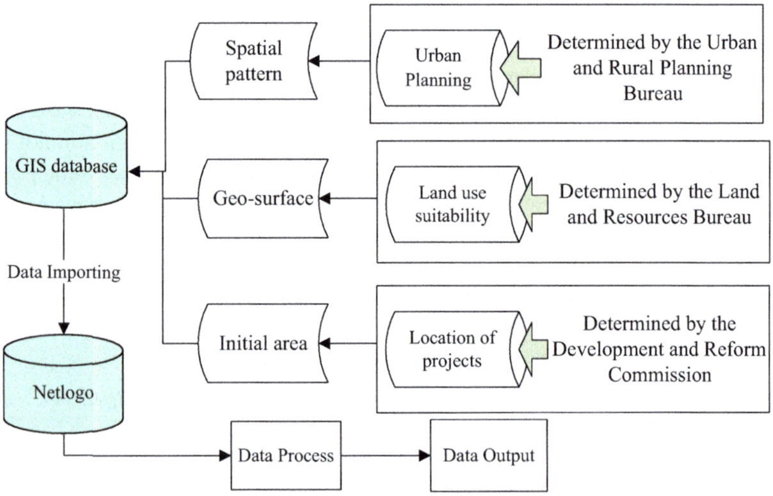

Fig. 3.2 The system framework

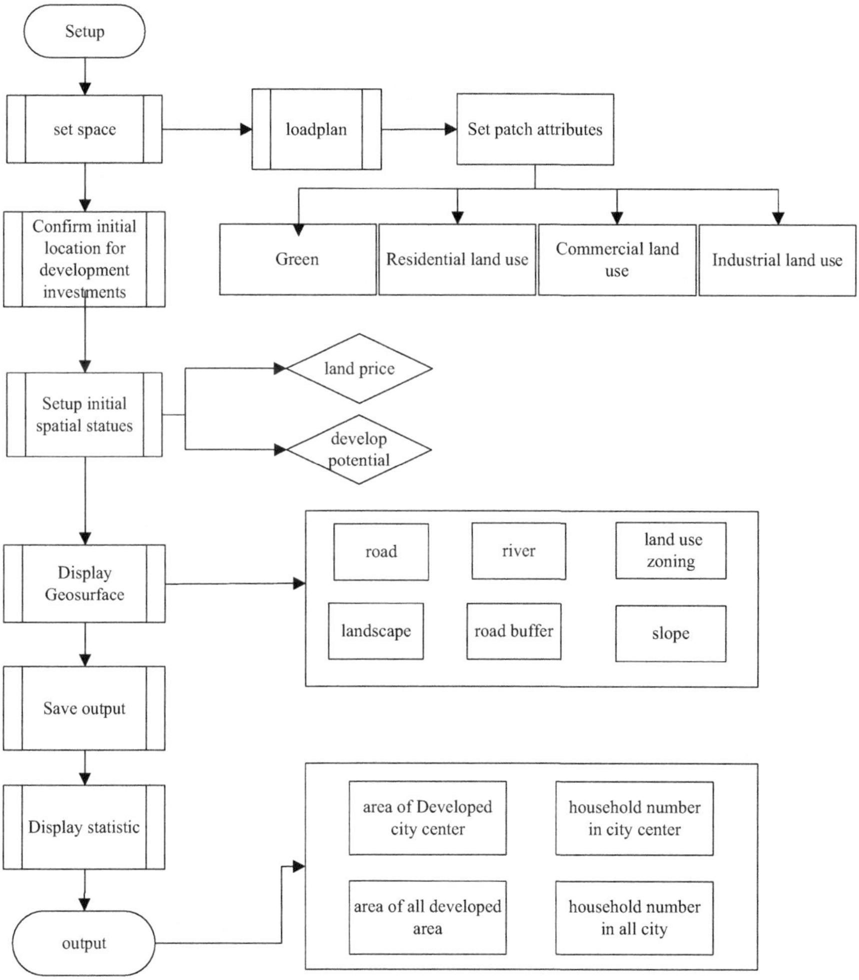

Fig. 3.3 Simulation flow of urban spatial dynamic

to the urban spatial dynamic simulation. For the urban spatial dynamic simulation, the CCA model is employed. During each simulation loop, the development potential of each land use cell will be calculated. As one of the incentives of urban spatial dynamic, locations of planned development investments will be set as the initial areas of urban spatial dynamic. Basing on the transition rule within CCA model, cell's development potential could be simulated in each simulation loop. The simulation results could be further used to statistical analysis in the system for calculating developed land cells in each land use zones or in the whole planning area. The analysis results can be observed by users through the system interface. Basing on the results of developed land use cells, the population aggregation could be simulated. Although we do not simulate the mobility of population inside the planning area, the scale of aggregated population could be simulated basing on the planning parameter of population density in different land use zones.

3.2.3 System Implementation and Results

We employed the system to simulate the spatial change and population aggregation in Chuandong area, China during its urban spatial dynamic process that affected by its spatial strategic plan. The Chuandong area belongs to Tongren city, which is in the northeast Guizhou Province, southwest China. The whole planning area is about 18.7 km^2 and local government was planning to develop a new city in that area. The land use zones of the new town are showed by Fig. 3.4, which divided the whole planning area into industrial area, landscape area, residential area, and commercial area. As shown as Fig. 3.4, the land use cells in T4, T5 and T6 areas are planned to be the center area of the new city. There will be real estate investments in this area. The land use zoning will be input to the simulation system and seemed as the institutional constraint of the CCA model.

Meanwhile, the topography of Tongren city will be input in the system as the spatial constraint of the CCA model (as shown as Fig. 3.5). As showed by the topography of Tongren city, the colored region is the planning area. All the land use zonings could only be conducted in the yellow area. The grey color cells area planned for industrial land use, the red area is for urban center, and orange cells are for residential.

The system is developed on the Netlogo platform too. Figure 3.6 shows the interface of the system. Through change the values in each slider, the different simulation views could be shown in the screen (black area), sliders are about parameters, global demands, development years and the coordinate of each developed cell (developed X and developed Y). There are still some settings for user choosing the different simulation results which would be showed in the screen. These settings are called choosers, as shown in Fig. 3.6, which named as spatial pattern, geosurface and environment assessment, respectively. The factors included in spatial pattern are land price, develop potential, priority for development and urban spatial dynamic. The factors displayed in geosurface are road, slope, plan, launching site and landscape. For environmental assessment, there are elements about possible pollution, possible energy consumption and possible domestic waste could be chosen.

According to the national economy and social development plan that issued by the committee of development and reform in local city, there are two major development investments for the new city. Figure 3.7 shows the sites of the projects, one is for developing an educational center and the other one is developing a business center. These sites will be seemed as seeds (the initial area) for urban spatial dynamic in the simulation model. In the meantime, public facility like highway is also planned (the gray lines in the Fig. 3.7). Basing on the different projects, we can design different scenarios for our simulation. For example, if we suppose the urban spatial dynamic is firstly start from the development of city center (including the educational center and business center), it means the cells of the center sites will have a high development potential at the initial stage. If we suppose the urban

Fig. 3.4 Spatial pattern of urban plan in Chuandong area

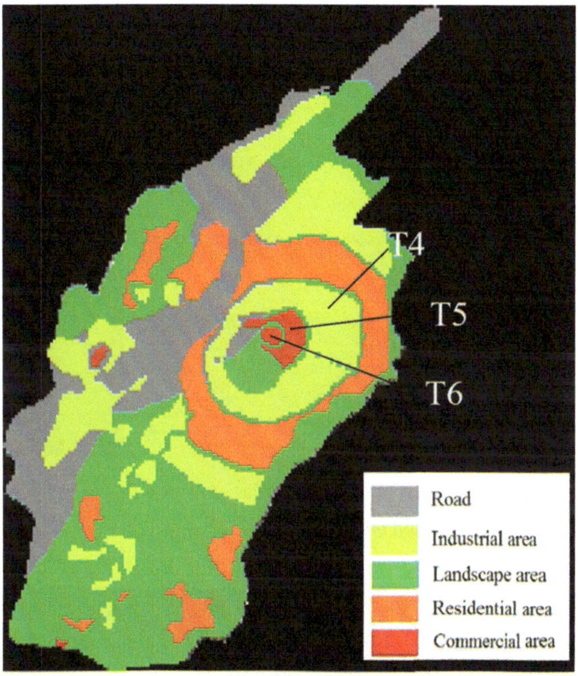

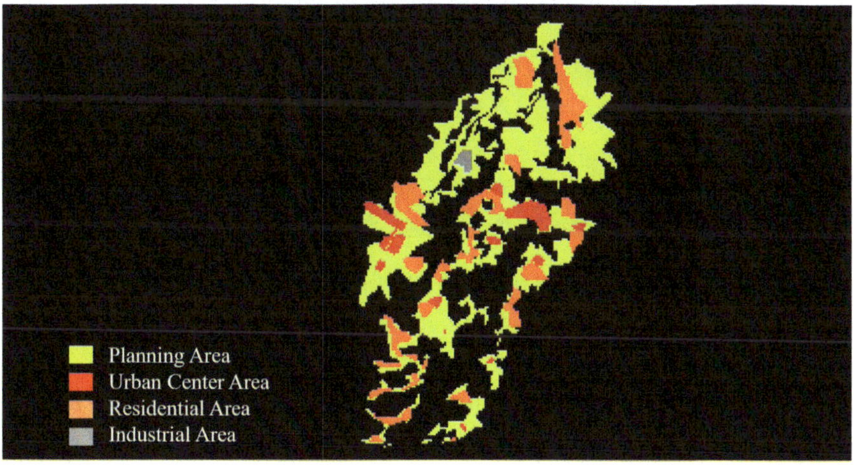

Fig. 3.5 The Topography of Tongren city

spatial dynamic will be both affected by city center development and highway development, the sites for planed highway and planned centers will have a high development potential till the initial stage. To answer whether different development strategies will lead to different spatial patterns, we simulated both the scenarios above.

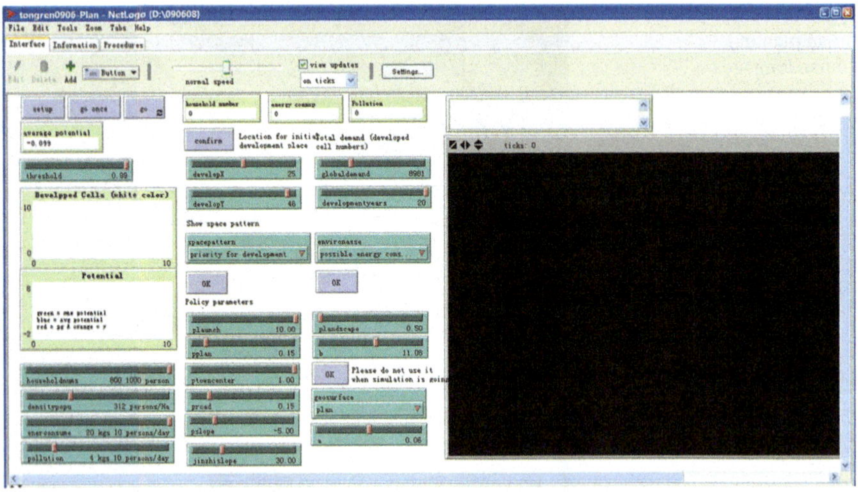

Fig. 3.6 the system interface

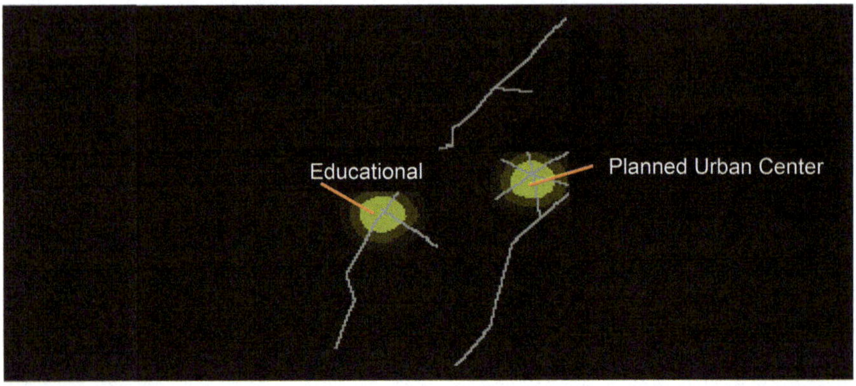

Fig. 3.7 the determined investment sites of local spatial strategic plan

Results for Scenario 1:

If suppose that the urban spatial dynamic is resulted from the city center development, the investment sites showed in Fig. 3.7 will be took as the initial areas of urban spatial dynamic. Dur to the local spatial strategic plan is divided into three stages, the short-term plan, mid-term plan and long-term plan, we designed the simulation to initial, middle and mature stages, respectively.

As shown as Fig. 3.9a, at initial stage developed land use cells are mostly centralized in the planned city centers, especially in the business center. It is resulted from the location sites of planned investment projects. Basing on the simulation results of developed land use cells, population aggregation is simulated to. But the simulation is taking the planned population density of developed cells as a basis. Darker the red color of the cell is and the higher population density of the cell is. The spatial pattern

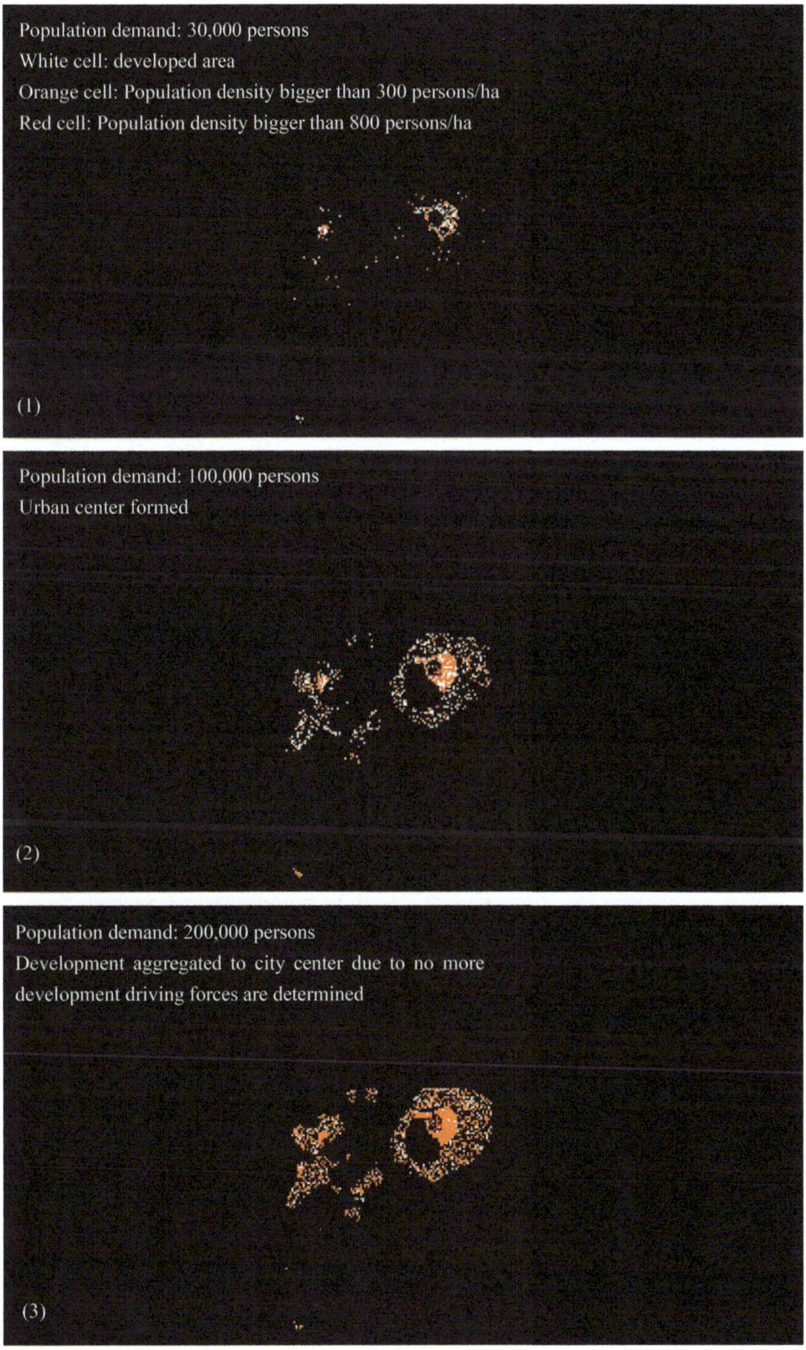

Fig. 3.8 (**a**) The simulation results of developed land use cells at initial stage of scenario 1. (**b**) The simulation results of developed land use cells at middle stage of scenario 1. (**c**) The simulation results of developed land use cells at mature stage of scenario 1

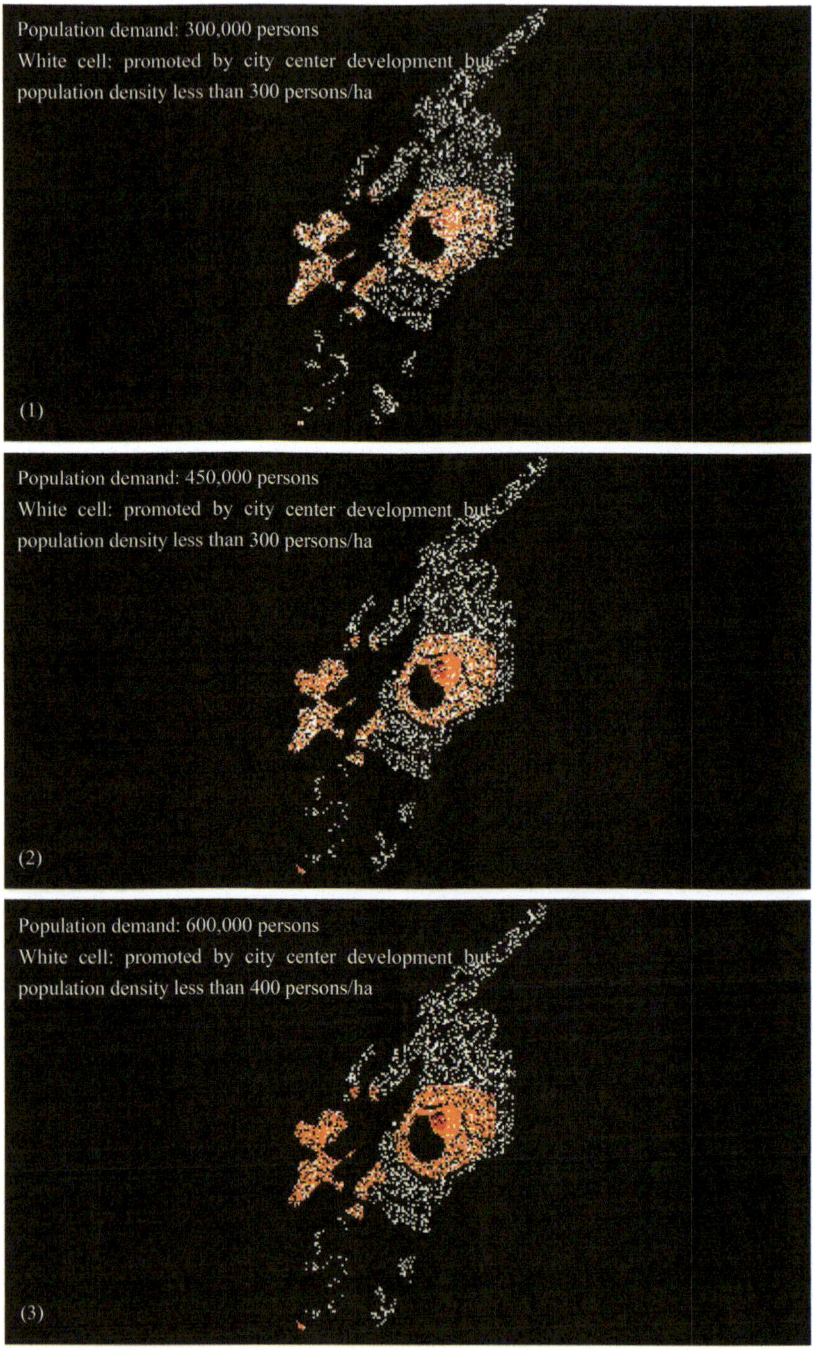

Fig. 3.8 (continued)

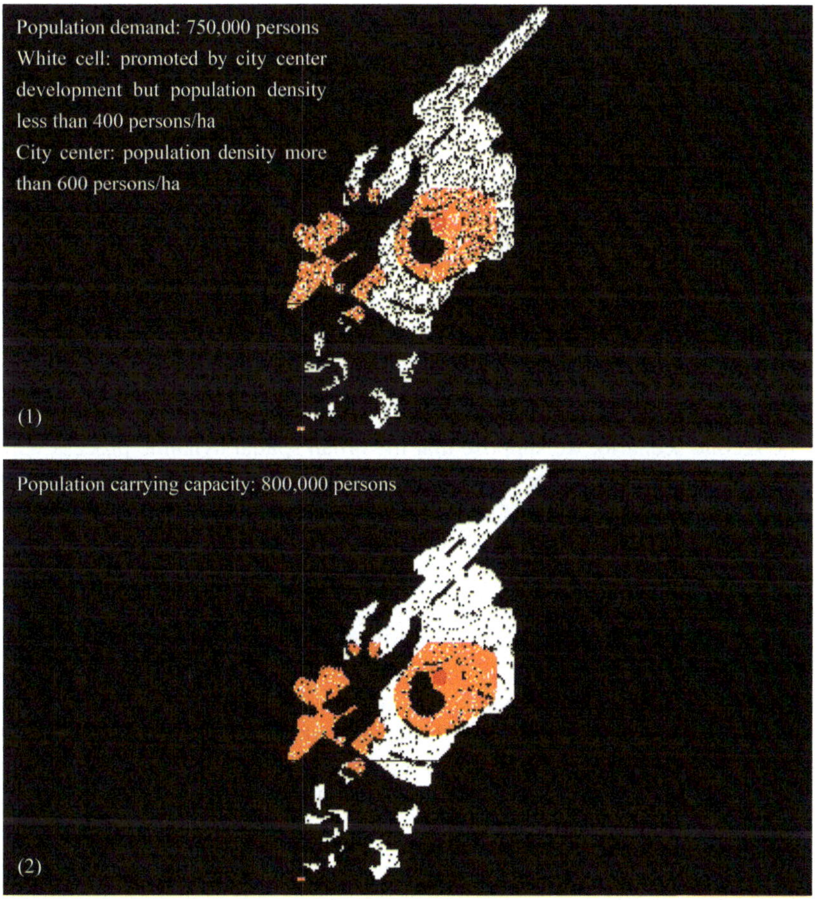

Fig. 3.8 (continued)

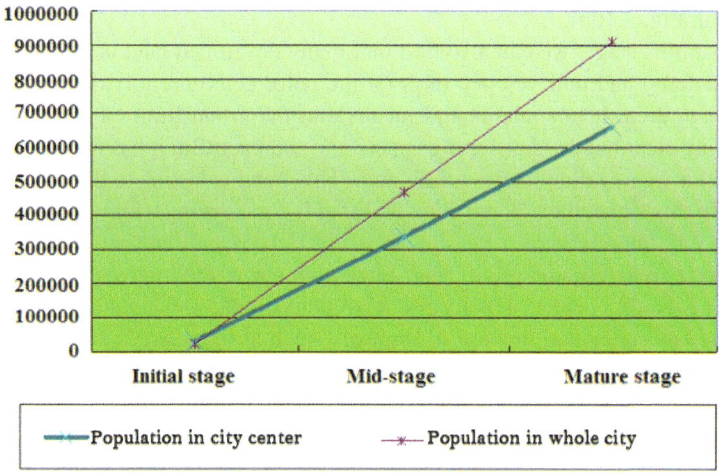

Fig. 3.9 Simulated households in scenario 1

of population distribution is that the colorful areas are mostly aggregated in business center area than those in the promoted areas. Till the end of initial stage, the population demand of research area is about 200,000 and mostly aggregated in city center areas due to limited driving forces of development (limited development projects).

Since the mid-stage of simulation, the developed land use cells gradually connected to the planned highway (as shown as Fig. 3.8b). This situation is further strengthened till the mature stage. The promoted area is growing rapidly and the scale of population demand increased to more than 800 thousand (see Figs. 3.8c and 3.9). Till this stage, the land use boundary of Chuandong new city is clear. It basically follows the master plan but the spatial pattern of developed cells is not compact, especially in the promoted area.

Results for Scenario 2:
In the second scenario, we suppose that the urban spatial dynamic process will be accelerated both by highway and city center development, but first start from the planned highway. In this case the planned highway (grey lines) is chose as initial areas for urban spatial dynamic (shown in Fig. 3.7). Except for the locations of initial area, the simulation parameters are the same with those of scenario 1.

At the initial stage of scenario 2, attracted by the planned highway the spatial pattern of developed land use cells follows a typical belt-form (as shown as Fig. 3.10a). Till the end of initial stage, accelerated by the development of planned urban center, although the spatial pattern is still a belt-form, the population gradually aggregate to the city center and the developed land use cells started to expend through the west to the east of the planning area. If we name this direction of urban expending as the west-east development axis, as the picture (1) shows in Fig. 3.10b, this axis will further strengthen and the spatial pattern of urban center becomes clear. While in this stage, except for the city center and highway, the other developed land use cells tend to be very dispersed. Thus, it is necessary to strengthen the construction of city center and planned highway. Through this effort the developed land use cells becoming more compact till the end of middle stage (as the picture (2) shown in Fig. 3.10b).

But in this stage, we can also find that many developed land use cells (white or colored cells) and undeveloped ones (white color cells) are mix distributed in the promotion area. It to some extend means that the urban form of planning area is becoming clear but the developing is not very compact. But it has not influenced on population aggregation to city center area. This trend is lasting to the end of mature stage. In this stage, the colored cells that stands for carrying a high population density become more compact (as shown as Fig. 3.10c) and the spatial pattern of the developed areas became quite clear. Since this stage, the highway is closely connected to the city center, thus local government should pay attention to reduce the negative environmental impacts of highway, such as using green belt to reduce the noisy and air pollution produced by traffics on the highway. Figure 3.11 shows the statistical result of aggregated population scale on developed land use cells.

In the case we introduced above, the driving forces on urban spatial dynamic are related to 4 aspects. Within the constraints, the land use suitability is an important factor to represent the suitability of land use cells on developed land use. Because of the scarcity of land resources and differentiations on location and economic attributes, land use suitability became a scientifical issue for spatial planning and suitable land use. In practice, land use suitability will be influenced by several factors, which can be basically divided into natural and economical factors. But in this case only slope has been considered (natural influencing factor), it is impossible to well reflect the carrying capacity and suitability of resources and environment. This is not a standard format in practice. If a simulation model is designed for planning practice, the evaluation process for land use suitability should follow the framework of planning practice. In the meantime, the neighborhood transition rule between

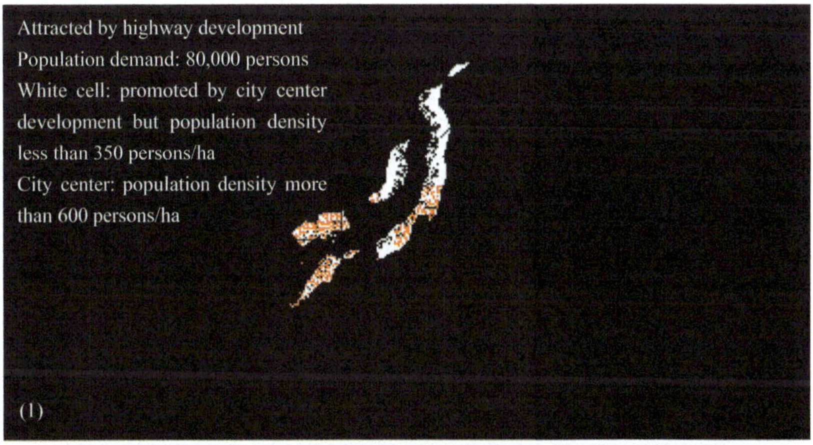

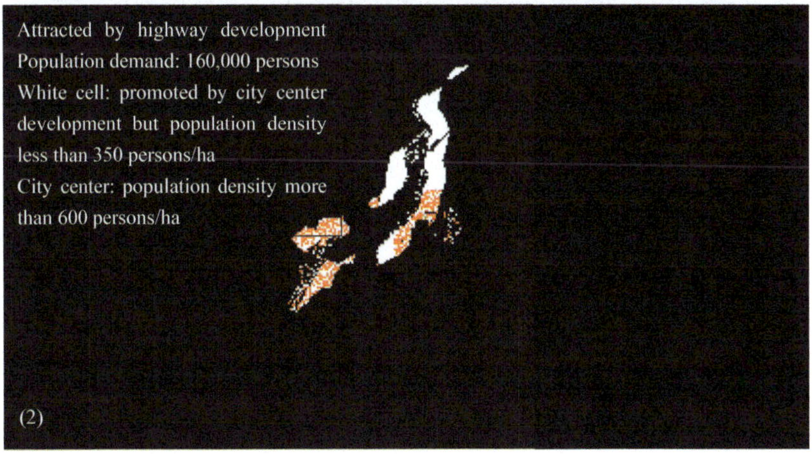

Fig. 3.10 (**a**) The simulation results of developed land use cells at initial stage of scenario 2. (**b**) The simulation results of developed land use cells at middle stage of scenario 2. (**c**) The simulation results of developed land use cells at mature stage of scenario 2

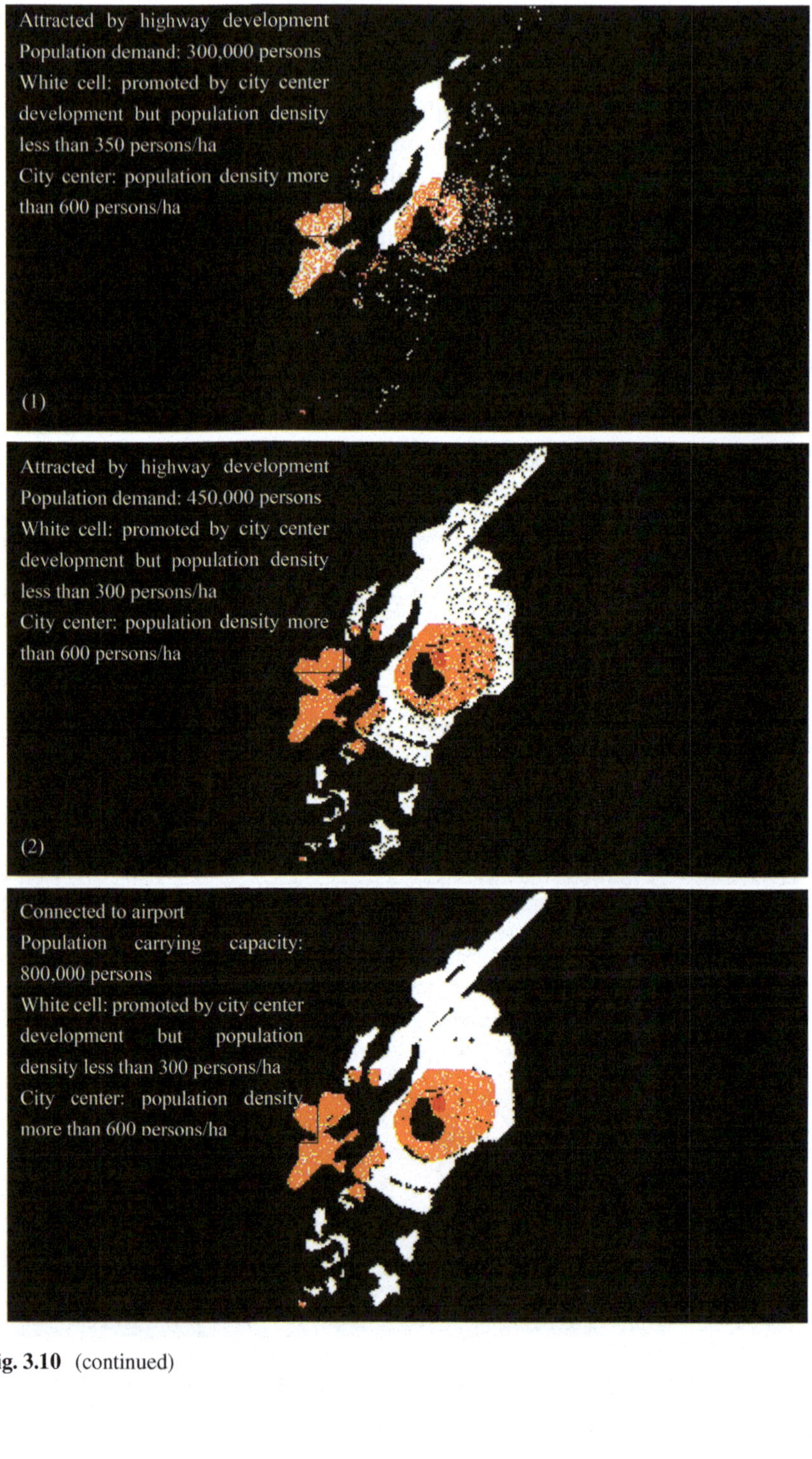

Fig. 3.10 (continued)

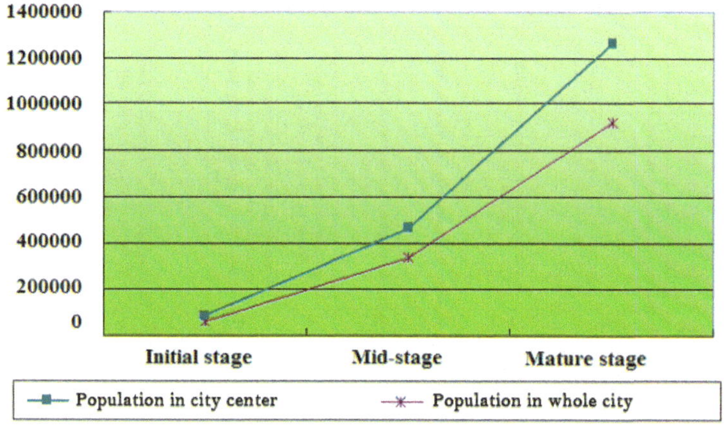

Fig. 3.11 Population in developed land use cells

neighborhood cells could be also improved from predefined rules to ANN. In the next subsection, we will introduce how to adapt the CCA model to territory spatial planning in China and new simulation technics, such as ANN.

3.3 CCA Model for Supporting Decision-Making of Territory Spatial Planning

3.3.1 The Establishing of Territory Spatial Planning and Basical Technical Needs

In 2019, the Central Committee of the Communist Party of China and the State Council issued "Several Opinions on Establishing a Territory Spatial Planning System and Supervising Implementation," which pointed out the integration of urban and rural planning, land use planning, and major functional area planning into territory spatial planning. As a basis of the planning, it is required to evaluate the Evaluation of Resource and Environmental Carrying Capacity (RECC) and the Territorial Development Suitability (TDS) (referred to as "double evaluation"). The double evaluation is served as a basis for scientifically and orderly arrangement of ecological, agricultural, urban and other functional zones. It is also the basis for delimiting ecological protection red lines, permanent basic farmland, urban development boundaries and other spatial control boundaries, such as sea area protection lines. The detailed structure and contents of territory spatial planning in China is showed in Fig. 3.12.

In this framework, the evaluation of TDS is one of the important technical methods to promote the implementation of this national strategy. At the practical level, the evaluation of the TDS mostly focuses on the analysis of the current status of

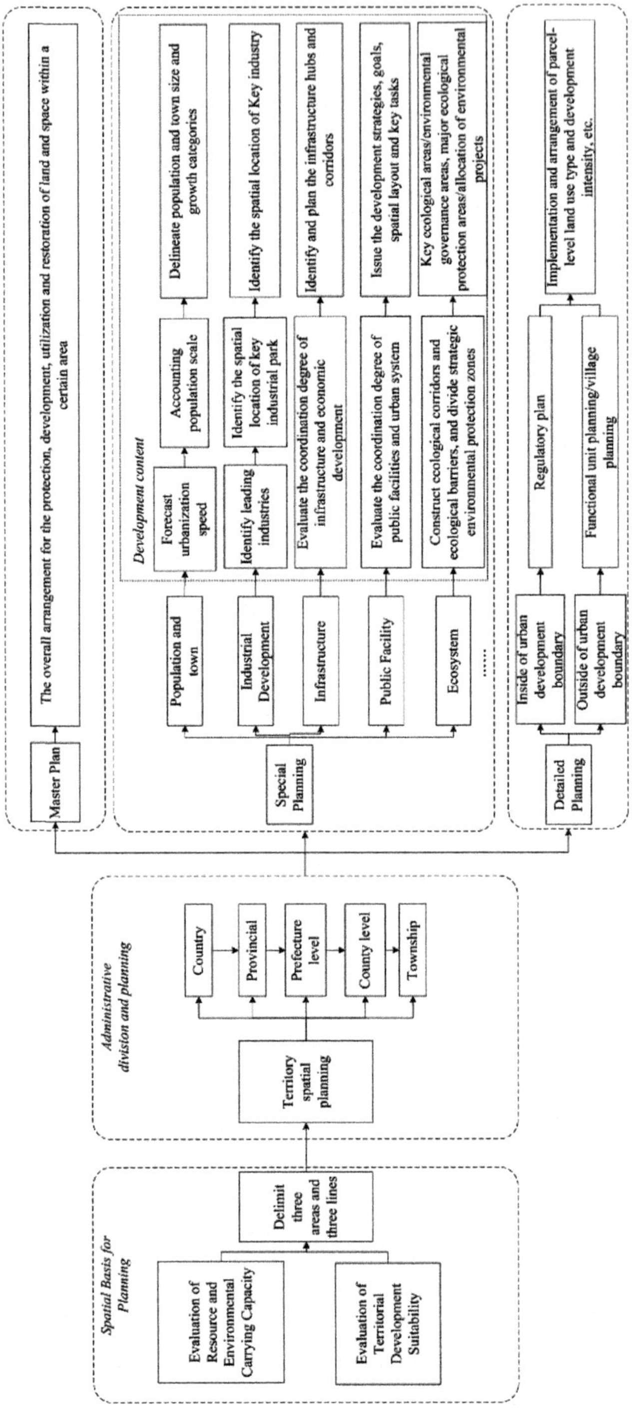

Fig. 3.12 The structure and content of territory spatial planning in China

territorial land resources (Yang 2016; Ji et al. 2019), and there is a gap between current evaluation and forecasting of future development trends. Modelling approach is a good attempt to mix this gap. For example, the European Commission considers the prediction of urban and regional land expansion when formulating its spatial planning policy, and predicts the future land use changes by integrating socio-economic factors, accessibility, land use suitability, controlled detailed planning, neighborhood relations and random variable simulations, and by setting several urban scenarios, adjusting model parameters accordingly, observing the effects of urban scenarios in different periods, and providing scientific support for the formulation of spatial planning and policies (Barredo et al. 2003, 2005; Lavalle et al. 2004). Japan's territorial and spatial planning system includes a land-use model simulation part, which takes the influencing factors such as spatial value (including land price and practicability), spatial distribution, competitiveness, multiplier effect, equilibrium, and dynamic methods into account, in order to provide scientific reference for the preparation of territorial and spatial planning (Aoyama 1989). But until now, the technical framework of Territory Spatial Planning in China has not included a modelling or simulation process.

In Sect. 3.3.2 we will give an example to introduce how to combine urban spatial dynamic simulation with issued technical framework of territory spatial planning, especially the double evaluation.

3.3.2 Coupling the "Double Evaluation" with CCA Simulation

The Jiangxi Province is chosen for case study. We will build a TDS evaluation index system based on the RECC of Jiangxi Province. Then CCA model will be combined with the evaluation system to simulate and predict the expansion of urban construction land resources at county level in Jiangxi Province till 2035. In this process the evaluation results of TDS is one of the constraints of CCA model.

In the process of compiling territory spatial planning, urban space is often delineated based on the results of the double evaluation, especially the suitability of urban construction land use. The double evaluation is used as an important basis for delimitation of the urban development boundary. It will play an inelastically constraint on the construction layout of strategic plan (representative example as master plan), detailed plan and special planning. However, due to the dynamic process of spatial and temporal nature of territory spatial development, it is necessary to couple the urban spatial dynamic simulation with the existing technical framework of territory spatial planning. The technical relationships between double evaluation and urban spatial dynamic simulation could be talked from 2 aspects, for the first one, in current technical framework the TDS is conducted under the spatial constraint of the RECC in a certain planning area, and the suitability will further effect on the abundance of future development land resources. For the second point, the urban spatial dynamic simulation forecasts the sprawl process of urban space, it helps us to find the urban spatial dynamic boundary. While we need to figure out that the simulated developed urban area cannot be simply seemed as the urban growth boundary of a research area. In this case, as shown as Fig. 3.13, the simulation is guided by the double evaluation, which is one of the constraints of urban growth. The simulation results will be further used to compare with the suitability evaluation results in order to final delineate the urban development boundary. The simulation results can also be used to support decision-making of rural high-quality development policies, and the layout of major projects.

By including planning constraints to CA model, the simulation results can reflect the effects of planning policy intervention, and ultimately support government decision-making (Ma et al. 2010). The simulation framework is showed in Fig. 3.14. For developing the simulation model, we referred to the FLUS model (Liu et al. 2017) to combine Markov chain, ANN and CCA method in this work. The detailed method is introduced in Sect. 3.3.3.

3.3.3 Methods for the Coupled Simulation Model

We use Markov chain to control the development scale of cellular automata at each simulation steps, thereby constraining the spatial development of cellular automata globally, and correcting the quantitative and spatial trajectory of cellular automata. The scale control we mentioned here is determined basing on the double evaluation. As a typical representative of this kind model, Markov model has been widely used (Xiao et al. 2011), and the formula is as follows:

$$X_{k+1} = X_k \times P \tag{3.3}$$

In the formula, X_k and X_{k+1} respectively represent the state vector of the trend analysis and prediction object at $t = k$ and $t = k + 1$, and P represents the one-step transition probability matrix.

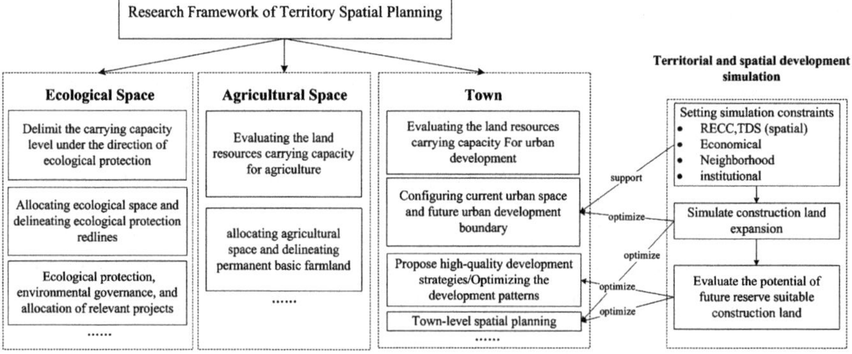

Fig. 3.13 Coupling the technology framework of territory spatial planning with urban spatial dynamic simulation

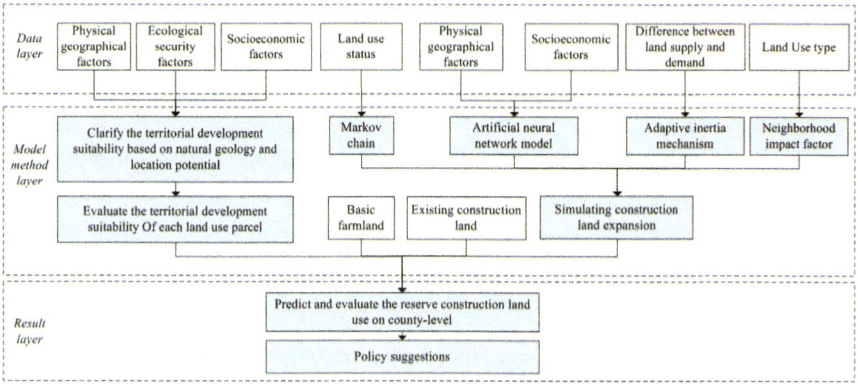

Fig. 3.14 The simulation framework for coupling double evaluation and CCA model

Artificial Neural Network (ANN) algorithm obtain the suitability probability of various land use types within the research scope from the first phase of land use data and various driving factors related to the natural environment and social economy (elevation, slope, traffic and urban influence). The algorithm is divided into 3 layers, the first layer is the input layer, the second layer is the hidden layer(like a black box), and the third layer is the output layer. The formula is as follows:

$$p(k,t,l) = \sum_i w_{i,t} \times sigmoid\left[net_i(k,l)\right] = \sum_i w_{i,t} \times \left[1 + e^{-net_i(k,l)}\right]^{-1} \tag{3.4}$$

In the formula, $p(k, t, l)$ represents the suitability probability of land use type t on cell k at time l; $w_{i,t}$, *sigmoid*( ) is the weight and activation function of the hidden layer and the output layer, spectively; it represents the response value of the raster k in the i-th hidden layer at time l; represents the response value of cell k in the i-th hidden layer at time l.

The FLUS model also proposes an adaptive inertial competition mechanism based on roulette selection, which can effectively cope with the uncertainty and complexity of land use changes under the combined influence of natural effects and human activities. The adaptive inertia coefficient is the main indicator that characterizes the adaptive inertial competition mechanism, and the formula is as follows:

$$Inertia_t^l = \begin{cases} Inertia_t^l & \text{if } \left|M_t^{l-2}\right| \le \left|M_t^{l-1}\right| \\ Inertia_t^l \times \dfrac{M_t^{l-2}}{M_t^{l-1}} & \text{if } 0 > M_t^{l-2} > M_t^{l-1} \\ Inertia_t^l \times \dfrac{M_t^{l-1}}{M_t^{l-2}} & \text{if } M_t^{l-1} > M_t^{l-2} > 0 \end{cases} \tag{3.5}$$

in which, M_t^{l-1} and M_t^{l-2} respectively represent the difference between the suitable cell number and the required cell number at simulation moment $l - 1$ and $l - 2$ for land use type t. The adaptive inertia coefficient of the land use type t at time l is expressed as $Inertia_t^l$.

The formula for cells diffusing developing information between automaton neighbors is showed below:

$$\Omega_{k,t}^l = \frac{\sum_{N\times N} con\left(c_p^{l-1} = t\right)}{N\times N-1} \times w_t \tag{3.6}$$

In the formula, $\sum_{N\times N} con\left(c_p^{l-1} = t\right)$ represents the total number of t type land use cells in the N-th order Moore neighborhood, w_t is the weight of neighborhood influence.

Integrating the suitability probability, adaptive inertia coefficient and neighborhood influence factor of each land use type, the overall land use conversion probability of each cell can be determined as follows:

$$TPS_{k,t}^l = p(k,t,l) \times \Omega_{k,t}^l \times Inertia_t^l \times \left(1 - sc_{c\to t}\right) \tag{3.7}$$

In the formula, the overall probability for cell k transform to land use type t at time l is $TPS_{k,t}^l$; $sc_{c\to t}$ is the cost of land use type c to t type, $1 - sc_{c\to t}$ represents the difficulty of the conversion.

3.3.4 Case Study Area

Jiangxi Province located in the southeastern China, on the south bank of the middle and lower reaches of the Yangtze River (Fig. 3.15). The province has a total area of 166,900 km^2 and a population of 46.661 million in 2019. It administrates 11 prefecture-level cities and 100 county-level cities. Jiangxi Province is surrounded by mountains from east, west, and south, with hills and plains staggered in the middle, and the Poyang Lake plain in the north. The land use structure of study area shows a character of high proportion of agricultural land use and low proportion of construction land resources use. As its important function on ecological protection (like Poyang lake ecological protection) and agricultural production in China, the urbanization in Jiangxi Province inevitably will be constrained by its functional zoning. Moreover, due to the constraint of topographical conditions, the suitable urban land use is quite scarce. Thus, it is very important to investigate the reserved land resources and its suitability and potential for ecological, urbanization and agriculture land use. The investigation results will be a key for making a reasonable territory spatial planning.

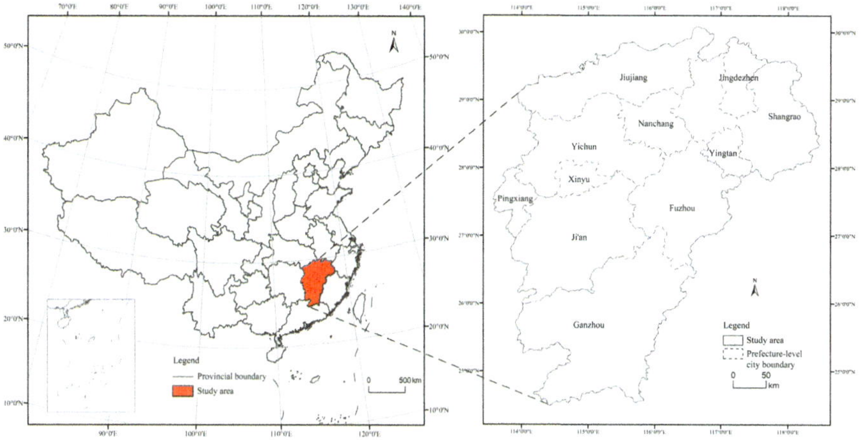

Fig. 3.15 Location map (left) and administrative division (right) of Jiangxi Province

This study respectively collected GIS data sources of 1:250,000 land-use map, basic farmland data, 1:250,000 digital elevation model (DEM) data, current and planned road network and railway data, geological hazard and scenic spots data in 1990, 2000, 2010, and 2015 in Jiangxi province, China. (here thanks for the supports of data source and research funding by the Innovation Center of the China Academy of Urban Planning and Design).

3.3.5 Recognizing the Spatial Constraints on Urban Spatial Dynamic

3.3.5.1 Building the Evaluation Index

Among various factors that affect land resources to be developed into construction land resources, we build the evaluation index from 2 aspects, which are natural and ecological security factors, socio-economic and location factors, respectively. The former is seemed as the inelastic influences on TDS and the latter is used to correct the influences. Why we use the socio-economic and location factors to correct the influences produced by natural and ecological security factors? In planning practice, we may meet situations that as the constraint threshold of natural resources we set, the land resources maybe not suitable for construction land resources development, while basing on technology progress or urbanization demands it developed in deed. For example, according to threshold we usually set for suitable urban construction land resources, the slope should not exceed 3 degree while in Chongqing city, China, affected by its topographic condition, land resources with a slope smaller than 3 degree is scarce. If we use the 3 degree as the threshold so that the natural resources of Chongqing city are not suitable for developing big scale city.

While as the central city in west China, although constraint by the slope urban land construction still be promoted by other factors for occupying its rapid urbanization. The other factors we mentioned here could be investments of developing projects, the location of city in the region, the urban function and developing target determined by planning policies or current engineering technologies (the acceptable slope for buildings with current technologies). These factors could be classified as socio-economic factors for urban spatial dynamic. As such, the effects of natural resources and ecological security on the TDS would be corrected by socio-economic factors but except for the inhibit developing land resources by law, such as the water bodies, geological hazard prone area, basic farmland, etc. Thus, in the TDS evaluation index, the socio-economic factors are elastic. Based on the selection of socio-economic and location factors as elastic correction adjustment indicators. Finally, 7 evaluation factors and 10 detailed indicators from 3 evaluation scopes are chosen to build the index (Table 3.1).

3.3.5.2 Evaluation of TDS

First, according to the natural geological conditions, the suitable grade of construction land resources is selected. Excluding the areas strictly prohibited by the corresponding laws and regulations for development, and the areas where the geological conditions are not suitable for development as construction land resources. Other areas are classified into four categories: suitable, relative suitable, logical suitable, and unsuitable (Table 3.2), according to the slope conditions and referring to the BP model to set thresholds. Secondly, the evaluation results are corrected according to socio-economic factors such as traffic and urban attractiveness. Among them, the selection of traffic factors is mainly based on the planned traffic system of Jiangxi Province. The influence of existing cities and towns mainly considers the

Table 3.1 Appraisal index of TDS evaluation in Jiangxi Province

Evaluation scopes	Evaluation factors	Detailed indicators
Physical geographical factors (inelastic constraints)	Terrain	Slope
	Waters	Water area
Ecological safety factors (inelastic constraints)	Areas with high incidence of geological disasters	Natural disaster-prone areas
	Scenic Zone	Protected area
	Natural reserve	Natural Reserve area
		Basic farmland
Socio-economic factors (elastic modification)	Traffic	Accessibility of current road
		Accessibility of current railway
		Accessibility of planned road
		Accessibility of planned railway
	Urban Attractiveness	Distance to built-up urban area

Table 3.2 Natural geological evaluation and location potential correction for Jiangxi Provincial TDS evaluation

Evaluation scopes	Evaluation factors and weights	Indicators and weights	Classification of evaluation results	Assignment
Natural geology (primary screening)	Slope	slope < 3°	Suitable	–
		3° < slope < 15°	Ralative suitable	–
		15° < slope < 25°	Logically suitable	–
		slope > 25°	Unsuitable	–
	Waters	Water area	Unsuitable	–
	Geological disaster	Natural disaster-prone area	Unsuitable	–
	Scenic Area	Protected area	Unsuitable	–
	Natural reserve	Natural reserve area	Unsuitable	–
		Basic farmland	Unsuitable	–
ccLocation potential (secondary correction)	Traffic (0.85)	Current railway accessibility (0.255)	<500 m Buffer	9
			500 ~ 1000 m Buffer	7
			1000 ~ 1500 m Buffer	5
			1500 ~ 2000 m Buffer	3
			>2000 m Buffer	1
		Planning railway accessibility (0.17)	<500 m Buffer	9
			500 ~ 1000 m Buffer	7
			1000 ~ 1500 m Buffer	5
			1500 ~ 2000 m Buffer	3
			>2000 m Buffer	1
		Current road accessibility (0.255)	<500 m Buffer	9
			500 ~ 1500 m Buffer	7
			>1500 Buffer	5
		Plan road accessibility (0.17)	<500 m Buffer	9
			500 ~ 1500 m Buffer	7
			>1500 Buffer	5
	Urban Attractiveness (0.15)	Urban built-up area (0.15)	<500 m Buffer	9
			500 ~ 1000 m Buffer	7
			1000 m ~ 2000 m Buffer	5
			2000 m ~ 3000 m Buffer	3
			>3000 m Buffer	1

neighborhood influence of existing urban land on future construction land resources development. The weight of each factor is mainly set with reference to the future strategic deployment of Jiangxi Province's construction and the characteristics of the transportation capacity of different transportation modes. According to the comprehensive index, the development potential index of each area is finally obtained, and it is divided into three levels: suitable, relative suitable, logically suitable. The suitability of construction land resources in Jiangxi Province was adjusted by combining the location potential grade and the natural environment suitability evaluation grade.

Finally, the scale and type of suitable construction land resources are obtained. The evaluation results show that the suitable construction land resources area in Jiangxi Province is 14591.06 km^2, accounting for 8.97% of the provincial area; the relatively suitable construction land resources area is 77918.83 km^2, accounting for 46.5% of the provincial area (Table 3.3). Although the total area of suitable and relative suitable land takes 55.47% of its provincial area, the current land use of suitable construction land resources mainly concentrates in the type of agricultural land use. It can be seen that the land resources available for development and construction in Jiangxi Province are very limited, and there is a problem of structural inadequacy. In the actual development process, to make good use of this kind of land resources, it is necessary to coordinate various relationships, such as occupation and compensation balance of cultivated land, basic farmland protection, and forest land protection.

According to the spatial differentiation characteristics of suitable construction land resources, the suitable construction land resources is basically distributed in the relatively flat Nanchang Basin, and generally presents a spatial distribution pattern of city clusters that centered with Nanchang City and two axes ('Fuliang County-Xinfeng County' suitable construction axis and the east-west suitable construction axis in the northern part of the province), three districts (suitable construction area of Nanchang City, suitable construction area of Ji'an-Ganzhou, suitable construction area of Yingtan-Shangrao-Jingdezhen), multi-node "(Fig. 3.16).

Table 3.3 Statistics of the suitable construction land resources in Jiangxi Province, China

Suitability level	Total area (km^2)	Area ratio	Main land use type	Suitable areas in main land use type (km^2)	Area ratio
Suitable	14591.06	8.97%	Construction land resources	1756.50	1.05%
			Agricultural land	12794.22	7.66%
Ralative suitable	77918.83	46.50%	Construction land resources	1308.43	0.78%
			Agricultural land	76502.88	45.82%
Logically suitable	27318.50	16.53%	Construction land resources	38.38	0.02%
			Agricultural land	27272.29	16.34%

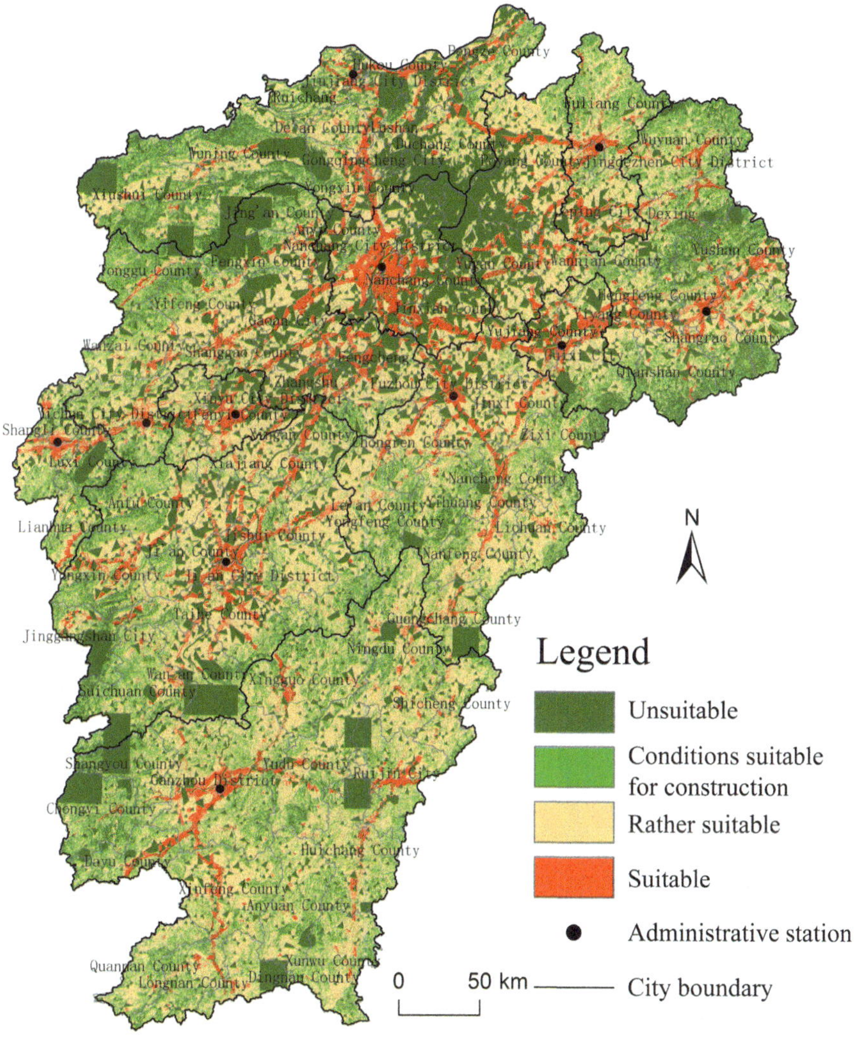

Fig. 3.16 Evaluation results of TDS in Jiangxi Province, China

3.3.6 Simulating Construction Land Resources Expansion During Urban Spatial Dynamic

Before the expansion simulation of construction land resources, we must verify the simulation accuracy. The verification method is to use the FLUS model to simulate and predict the distribution of various types of land in Jiangxi Province in 2010 based on the distribution of various types of land in Jiangxi Province in 1990 and 2000. The accuracy of the model was verified by comparing the actual distribution of various types of land in Jiangxi Province in 2010. On the one hand, the Kappa coefficient is introduced, and the overall accuracy of the calculation can reach

0.9211, the Kappa coefficient is 0.8536, and the simulation accuracy is high. On the other hand, the Fom coefficient proposed by Pontius (Pontius et al. 2008) is used to evaluate the accuracy of urban spatial dynamic. The calculated Fom coefficient is 0.14, and the simulation accuracy is also high. Taking the predicted construction land resources scale of Jiangxi Province in 2035 as the target, run the FLUS model to generate various land distributions in Jiangxi Province in 2035 through iteration (Fig. 3.17).

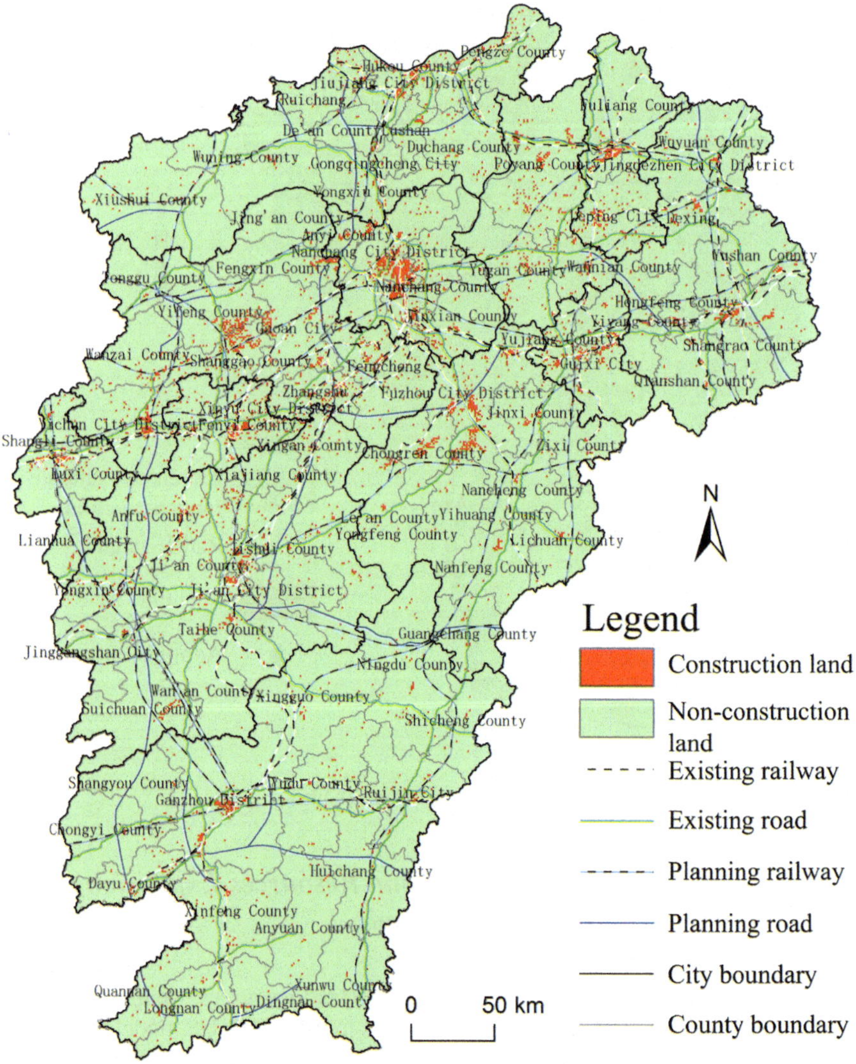

Fig. 3.17 Simulation of expansion of construction land resources in Jiangxi Province in 2035

The simulation results show that the construction land resources area of Jiangxi Province will be 4932 km^2 in 2035, an increase of 24.67% over 2010, showing that Nanchang is the core city node, and Fuzhou, Yichun, and Shangrao surrounding Nanchang are the main expansion areas. The sprawl pattern shows a "Point-to-pole" mode. Among them, the construction land resources of Fuzhou City mainly expands to the urban area and Chongren County; the construction land resources of Yichun City and Shangrao City breaks through the urban area, showing an enclave expansion character; the construction land resources of Yichun City mainly expands to Shanggao County and Gaoan City and the construction land resources of Shangrao City is mainly expanded to Poyang County.

The expansion of construction land resources is a corresponding result of the agglomeration of population and industry. It shows the necessity of planning and developing a corresponding scale of infrastructure to accelerate the flow of productive factors within the expanded urban areas. In this case, we further investigated the current and planned transportation infrastructures in the areas that large expansion may happen showed by the simulation from 2010 to 2035. These areas are mainly Poyang County, Linchuan District, Shanggao County, Gaoan City, Yifeng County, Yushui District, Fengcheng City, Zhangshu City, Anfu County and Leping City). The analysis found that the construction land resources may increase a lot in the middle of Poyang Lake, the southwest of Linchuan District, and the borders of Yifeng County, Gaoan City, and Shanggao County, but there the current and planned railways, highways and other transportation infrastructure are less. It should be improved in practice.

The expansion of construction land resources in the southern cities of Jiangxi Province is generally slow. This simulation result may relate to the wide coverage of mountains and hills in the south. The plain area in this region is rare. In the meantime, a large area in the south is restricted by the ecological protection red line. In order to further explore the spatial coupling relationship between construction land resources and suitable construction land resources in Jiangxi Province after expansion, we also calculate the reserved suitable construction land resources in Jiangxi Province from the county level. Reserve construction land resources is called usable land resources or land resource abundance in China's "Provincial Technical Regulations for the Division of Main Functional Zones", which can objectively reflect the carrying capacity of land resources for different regions in future population aggregation, industrialization and urbanization, and the scale of economic development (Dang et al. 2014). Reserved suitable construction land resources focuses on evaluating the abundance and shortage of reserved suitable construction land resources from the perspective of absolute quantity of land resources.

The formula for calculating reserved suitable construction land resources is as follows:

[Reserved suitable construction land resources] = [Suitable construction land resources area] – [Existing construction land resources area] – [Basic farmland area].

[Existing construction land resources area] = [Urban land area] + [Rural residential land area] + [Independent industrial and mining land area] + [Transportation land area] + [Special land area] + [Water conservancy facility construction land resources area]

Assumed that the area of suitable construction land resources and basic farmland in Jiangxi Province will remain unchanged by 2035, the above formula calculates that the area of reserved suitable construction land resources in Jiangxi Province in 2035 is 9372.64 km^2, accounting for 5.62% of the total provincial land area (Table 3.4).

On the prefecture-level, the total area of reserved suitable construction land resources in Ganzhou, Shangrao, Ji'an and Nanchang is more than 1000 km^2. While most reserved suitable construction land resources concentrate in Nanchang, which takes 14.27% of the total amount and takes up nearly 10% of its urban area. The ratio in Ganzhou City, Shangrao City and Ji'an City occupies between 3% and 7% of the city area. It can be seen that the suitable construction land resources in Nanchang City will be relatively abundant land by 2035. It means that its potential for territory space development is relatively high. It should belong to its terrain, which produces a relatively slow slope in that region. Ganzhou City, Shangrao City, and Ji'an City have a low potential for usable land resources by 2035. In particularly, the per capita usable land resources are relatively small, resulting in a phenomenon of artificial high. It mainly due to the urban areas of the three cities are big while dominated by mountains and hills. In contrast, although the reserved suitable construction land resources area in Yingtan City is relatively small, the reserved suitable construction land resources area accounts for 11.15% of the city area. This value is just smaller than Nanchang City and ranked number two in Jiangxi Province on prefectural level. It shows that Yingtan City also has a relative high potential on territory space development.

According to the classification threshold of usable land resources in the "Technical Regulations for the Division of Main Functional Areas at the Provincial

Table 3.4 Potential of suitable land for construction in various cities in Jiangxi Province in 2035

Prefecture-level city name	Reserve construction land resources area (km^2)	Reserve construction land resources accounts for the proportion of city area (%)
Nanchang	1026.78	14.27%
Jingdezhen	426.66	8.13%
Pingxiang	297.24	7.73%
Jiujiang	796.22	4.17%
Xinyu	266.38	8.42%
Yingtan	397.50	11.15%
Ganzhou	1517.22	3.86%
Ji'an	1401.31	5.54%
Yichun	950.16	5.09%
Fuzhou	866.60	4.61%
Shangrao	1426.90	6.28%
Total	9372.64	5.62%

Level" and the characteristics of the prediction and research in this work, the abundance of Jiangxi Province's reserved suitable construction land resources is divided into 6 grades on county level, which are rich (more than 320 km^2), rather rich (150–320 km^2), medium (100–150 km^2), rather lack (50–100 km^2), lack (0–50 km^2), and overburden (less than 0) (Fig. 3.18).

As a summary, the spatial distribution of reserved suitable construction land resources in Jiangxi Province has the following characteristics: the amount of reserved suitable construction land resources in the northeastern region is more than in the southwest region, in the central area is more than in the outside region, and in

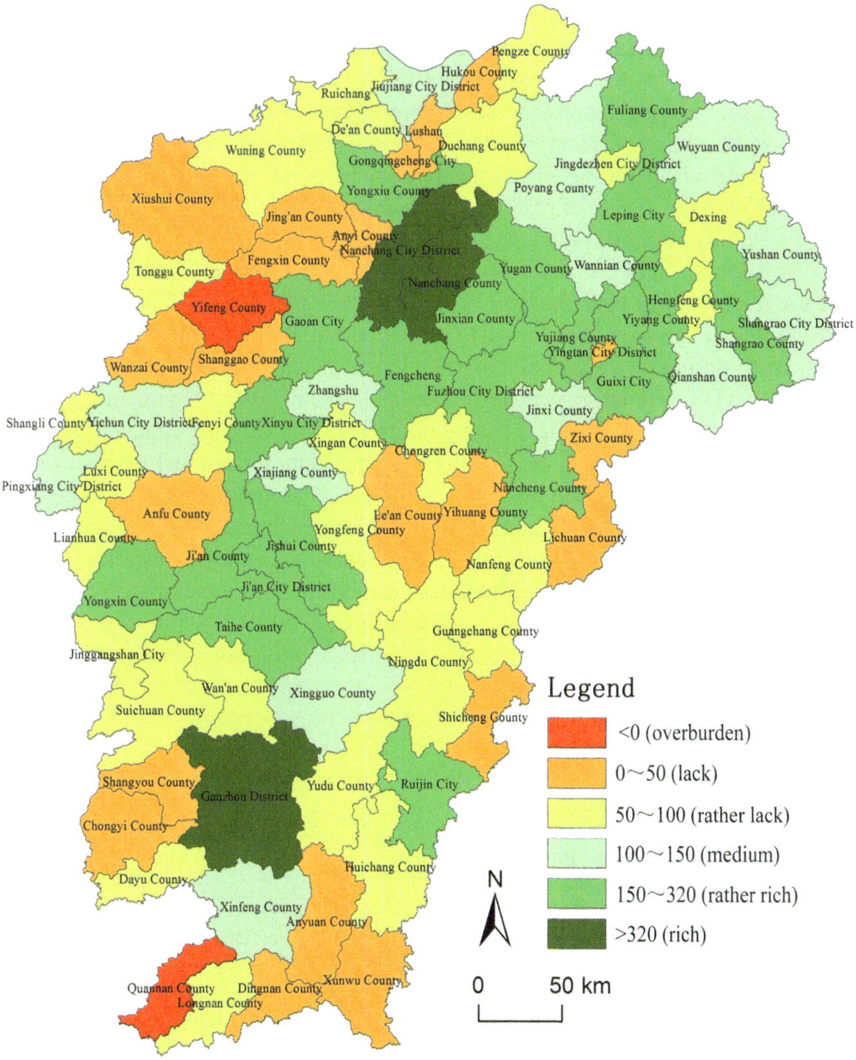

Fig. 3.18 The abundance grade of reserved suitable construction land in Jiangxi province till 2035

the plain basin area is more than the mountain and hilly region. From the perspective of the abundant grade of reserved suitable construction land resources, 56.98% of county-level units will be lack or even overburden till 2035.

3.4 Discussion

In this chapter we introduced how to simulate urban spatial dynamic. Accompanying with the development of information technology, the traditional CA model has been developed to combined CCA and ANN, it makes the simulation can better serve planning practice. Despite of the CCA model has great similarity with spatial planning practice due to it considers different driving force or we call them constraints on urban spatial dynamic, the spatial constraints is still limited in reflecting the carrying capacity of local land resources. Urban spatial dynamic is a process that taking history as a basis will facing to the future, this character requires the simulation not only can reflect the historical mechanism of urban spatial dynamic but also consider the future effects of planning intervention. Such as the development potential (developing possibility) defused between cells (named as neighborhood effects) or socio-economic effects, as the second example we introduced in this chapter, we can use the historical land use date to recognize the neighborhood effects between cells, it makes the simulation model can better adapt to the real land use change mechanism during urban spatial dynamic. In the meantime, by combining Markov chain with TDS evaluation, we can determine the scale of future construction land and constraint land development basing on its TDS and carrying capacity. This is a typical planning awareness. This point is very important for a model whether can be used to planning practice. Till this stage, although cells cannot move and make decisions, the spatial change and construction land development during urban spatial dynamic could be simulated and follows a planning awareness. Since last chapter we will further introduce how will human spatial behaviors interact with urban space.

References

Aoyama Y. A historical review of transport and land-use models in Japan. Transportation Research Part A: General, 1989, vol. 23(1), pp. 53–61.

Barredo, J.I., Kasanko, M., McCormick, N., Lavalle, C., 2003. Modelling dynamic spatial processes: simulation of urban future scenarios through cellular automata, Landscape and Urban Planning, vol. 64(3), pp. 145–160.

Barredo J. I., Petrov L; Sagris V; et al., 2005. Towards an integrated scenario approach for spatial planning and natural hazards mitigation. European Communities, DG-JRC, Ispra: EUR, 21900.

Batty, M., Couclelis, H., Eichen, M., 1997. Urban systems as cellular automata, Environment and planning B: planning and design, vol. 24, pp. 159–164.

Batty, M., Xie, Y., Sun, Z., 1999. Modeling urban dynamics through GIS-based cellular automata. Computers Environment & Urban Systems, vol. 23, pp. 205–233.

Brown D.G., Robinson D.T., 2006. Effects of Heterogeneity in Residential Preferences on an Agent-Based Model of Urban Sprawl. Ecology and Society, 11(1), 46–67.
Clarke, K.C., Gaydos, L.J., 1998. Loose-coupling of a cellular automaton model and GIS: long-term growth prediction for the San Francisco and Washington/Baltimore, International Journal of Geographic Information Science, vol. 12, pp. 699–714.
Cao, K., Huang, B., Li, M., Li, W., 2014. Calibrating a cellular automata model for understanding rural-urban land conversion: a Pareto front-based multi-objective optimization approach. International Journal of Geographic Information Science, vol. 28, pp. 1028–1046.
Cao, M., Bennett, S.J., Shen, Q., Xu, R., 2016. A bat-inspired approach to define transition rules for a cellular automaton model used to simulate urban expansion. International Journal of Geographic Information Science, vol. 30, pp. 1–19.
Dang L.J., Xu Y; Tang Q., et al., 2014. The potential and spatial distribution of reserve suitable construction land along the Xijiang River in Guangxi. Journal of Natural Resources, 29(3): 387–397.
Fang, S., Gertner, G., Sun, Z., Anderson, A., 2005. The impact of interactions in spatial simulation of the dynamics of urban sprawl, Landscape and Urban Planning, vol. 73, pp. 294–306
Fontaine C.M., Rounsevell M.D.A., 2009. An agent-based approach to model future residential pressure on a regional landscape. Landscape Ecological, 24, pp. 1237–1254.
Han J., Yoshitsugu H., Cao X., Hidefumi I., 2009. Application of an integrated system dynamics and cellular automata model for urban growth assessment: A case study of Shanghai, China. Landscape and Urban Planning, vol. 91, pp. 133–141.
Han H.R., Yang C.F., Song J.P., et al., 2015. Scenario simulation and the prediction of land use and land cover change in Beijing, China. Sustainability, 7(4), 4260–4279.
Ji X. P., Huang X. J., Chen Y., et al., 2019. Comprehensive suitability evaluation of spatial development and construction land in the perspective of land-ocean co-ordination: A case study of Liaoning province, China. Journal of Natural Resources, vol. 34(3), pp. 451–463.
Lavalle C., Barredo J. I., McCormick N., et al., 2004. The MOLAND model for urban and regional growth forecast. A tool for the definition of sustainable development paths. Ispra: EC-DG-Joint Research Centre, Institute for Environment and Sustainability-Land Management Unit, European Communities.
Liu X P; Liang X; Li X; et al. A future land use simulation model (FLUS) for simulating multiple land use scenarios by coupling human and natural effects. Landscape and Urban Planning, 2017, 168 (94–116).
Li, X., Yeh, A.G.O., 2000. Modeling sustainable urban development by the integration of constrained cellular automata and GIS, International Journal of Geographic Information Science, vol. 14, pp. 131–152.
Li, X., Lin, J., Chen, Y., Liu, X., Ai, B., 2013. Calibrating cellular automata based on landscape metrics by using genetic algorithms. International Journal of Geographic Information Science, vol. 27, pp. 594–613.
Long Y, Shen, Z, Du L, Mao Q, Gao Z, 2008. BUDEM: an urban growth simulation model using CA for Beijing metropolitan area, Proc. SPIE, Vol. 7143.
Long Y, Shen Z, Mao Q, Dang A, 2009. Form scenario analysis using constrained CA, Proceedings of Computers in Urban Planning and Urban Management.
Marwa Waseem A. Halmy, Paul E. Gessler, Jeffrey A. Hicke, Boshra B. Salema, 2015. Land use/land cover change detection and prediction in the north-western coastal desert of Egypt using Markov-CA. Applied Geography, vol. 63, pp. 101–112.
Munshi, T., Zuidgeest M., Brussel M., Maarseveen M.V., 2014. Logistic regression and cellular automata-based modelling of retail, commercial and residential development in the city of Ahmedabad, India. Cities, vol. 39, pp. 68–86.
National Bureau of Statistics, 1996–2005. China Statistical Yearbook, Beijing: China Statistics Press, [中国统计年鉴].
Poelmans, L., Rompaey, A.V., 2010. Complexity and performance of urban expansion models. Computers Environment & Urban Systems, vol. 34(1), pp. 17–27.

Pontius, R. G., Boersma, W., Castella, J. C., Clarke, K., de Nijs, T., Dietzel, C., ... & Verburg, P. H. 2008. Comparing the input, output, and validation maps for several models of land change. The Annals of Regional Science, 42(1), 11–37.

Saa R., Liu W., 2022. Predicting future urban growth scenarios and potential urban flood exposure using Artificial Neural Network-Markov Chain model in Miami Metropolitan Area. Land Use Policy 114.

Shen Z, Kawakami M, 2008. Geo-simulation model using geographic automata for simulating land use patterns in urban partitions, Environment and planning B: planning and design, vol. 36, pp. 802–823.

Torrens P.M., Benenson I., 2005. Geographic Automata Systems. International Journal of Geographical Information Science, vol. 19(04), pp. 385–412.

Torrens P.M., 2007. A Geographic automata model of residential mobility. Environment and Planning B: Planning and Design, Vol. 34, pp. 200–222.

Ward D.P., Murray A.T., Phinn S.R.A., 2000. Stochastically Constrained Cellular Model of Urban Growth. Computers. Environment and Urban System, vol. 24(6), pp.539–558.

White R., Engelen G., 1997. Cellular automata as the basis of integrated dynamic regional modelling. Environment and Planning B: Planning and Design, vol. 24(2), pp. 235–246.

Wu F., 1998. Simland: a prototype to simulate land conversion through the integrated GIS and CA with AHP-derived transition rules, International Journal of Geographic Information Science. vol. 12(1), pp. 63–82.

Xiao X., Li Y. F., Zhu X. D., et al., 2011. Markov model based on driving forces of land use change and its application. Soil, vol. 43(5), pp. 822–827.

Yang Z.S., 2016. Land Suitability Evaluation for Urban Construction and Its Application in Mountainous Areas: A Case Study in Dehong Dai-Jingpo Autonomous Prefecture, Yunnan Province. Journal of Natural Resources, vol. 31(1), pp. 64–76.

Ma Y., Shen Z.J., Long Y., Kawakami M., Wang K., Suzuki K., 2010. Urban Growth Simulation for Spatial Strategic Plan of Chuangdong Area, China, published by the 18th International Conference on GeoInformatics.

Chapter 4
Agent-Based Simulation of Household Residential Relocation and Decision-Making Support of Downtown Revitalization

4.1 Introduction

In last Chapter, we introduced how to simulate the urban growth process effected by spatial strategic plan. In the model, population aggregation has been simulated basing on the planning condition of population scale and density. While when population aggregate in urban area, they usually firstly need to make a residential location. Thus, in this chapter we will introduce how household agents make residential location and relocation choices. We will also use a case study to introduce how we employed ABM in simulating household residential relocation in order to forecast the policy effectiveness on downtown regeneration.

Residential mobility is always a hot concern by researchers in urban study area. Researchers managed to explain the interactive process of household residential mobility, land use changes and housing markets during urban growth (Mendonca et al. 2020; Haase et al. 2010). In this period, the residential mobility always stands for sprawling settlement during urbanization. In the meantime, because that the growing urban space and sprawling settlement may lead to long commuting distance, researches regarding to the interactive process between residential location choice and transportation development are carried out (Wang et al. 2020; Klinger 2017; Gerber et al. 2017). Through these researches, the correlate factors with residential location choices have been gradually revealed and hereby, the spatial patterns of residential mobility have been widely investigated by researcher (Tannier et al. 2020; Li et al. 2020), some spatial patterns such as segregation and its causes have been concerned too (Ardestani et al. 2018; Fossett 2015; Bruch 2014).

As showed by European experience in urban development, generally there are three stages for an urban development, which are respectively urbanization, suburbanization, and urban decline (Klaassen and Paelinck 1979). Since urbanization has been replaced by suburbanization or even urban decline in many developed countries, research focuses in sprawl settlement is changing to urban shrinkage (Rieniets

Y. Ma, Z. Shen, *Strategic Spatial Planning Support System for Sustainable Development*, Advances in Geographic Information Science,
https://doi.org/10.1007/978-3-031-07543-8_4

2009). Thus, researchers in the field of household residential mobility began to find the planning interventions to accelerate household relocation to urban center area in order to make our urban form more compact (Jordana et al. 2014; Ma et al. 2013; Gaube and Remesch 2013; Lee et al. 2010). Through these researches we can find that no matter researching on different spatial pattern of residence, sprawl settlement or planning approach for compact urban form, researchers all concentrate on reveal the interactive between human residential behavior and spatial feedbacks. The main difference is whether we concern the policy intervention during the above process. In this chapter we will use a case study to talk how to develop such an ABM for planning support of downtown regeneration.

4.2 Existing Researches for Household Residential Location Choice

It is assumed that residential location is closely tied to one's commuting mode (Cao et al. 2009). Thus, regarding to individual's preference on residence, there are many experiments focusing on investigating the interactive relationship between transport characteristics and the characters of residential location (Fatmi and Habib 2018; Chen et al. 2008; Molin and Timmermans 2003). The statistical results in such researches could apply us with a good referee to the influencing factors for human decision making on location choice. Such factors could be the social statues of residents, the density of residential environment, distance from work place, the amenities for parking or public transport, ect. One of the representative models in this area is the Land Use and Transport Interaction (LUTI) models. At first, the LUTI simulates population behaviors on commuting or location choices by seeming human as an aggregated category. Since agent-based simulation widely been applied in urban modeling, it is also employed in developing LUTI models (Waddell et al. 2003). This kind LUTI model took whole city as research target to simulate and reflect inhabitant personal behavior on residential choices interacted with land use and urban transportations. This is a quite different from former works in this field. It shows the possibility in urban simulation that develop a model to simulate complex urban system (Benenson 1998; Miller et al. 2004). Another representative model in this field is UrbanSim (Noth et al. 2003), which was developed as an ABM for simulating activities of different actors and urban development processes within an urban environment.

But all these researches have not taking policy intervention into account so that it is impossible for planning support of targeted planning policies. Thus, how to utilize ABM to policy simulation has become a new concern by researchers. By developing ABMs, it is possible to simulate and forecast the future socio-ecological consequences resulted from targeted land use policy (Lee et al. 2010) or forecasting the urban development process by input multi-scenarios of development policies (Berger et al. 2006), and even the spatial pattern of urban population and their demand of different public facilities (MA et al. 2016). In the meantime, ABM has also been employed in housing market for simulating land development and housing price (Ettema 2011; Dawn and Filatova 2008; Diappi and Bolchi 2006). In these researches we can find a similarity in simulating household residential location choices, which is households make location choices basing on the evaluation of dwellings between different location alternatives (Moreno and Moeckel 2018). This logic is widely accepted by researcher for simulating the decision of household agents on location choices. In this chapter, our model is also developed on this logic and household agent will evaluate the utility of different locations. But how to consider the policy intervention in households' utility evaluation is one of the problems we are going to solve in this chapter.

Meanwhile, we need to distinguish households' behaviors in residential location and relocation. For the former, the utility model can be used to help household agents to make choices. While for the latter we still need to figure out when will household agents make relocations., although various methods could be used to investigate when will a household face to a residential relocation, a consensus is basically gained that there are two kinds of occasions household would show a preference for relocation, the first one is resulted from unsatisfied with current location and the other one may resulted from the change of one's lifecycle stage (López-Ospina et al. 2016; Fontaine and Rounsevell 2009). The former could be defined as adjustment move and the latter is incentive move (induced move). In this Chapter we will consider both the two kinds of reasons in our agent-based relocation model. In order to support policy decision making, the model will also simulate the interactive process between household agents regarding to a residential promoting policy in case study area. The model is supposed to reflect the residential behaviors of households by different age groups. Basing on the simulation results, we are going to investigate the effectiveness of the residential promoting policy on downtown revitalization.

4.3 Approach of Agent-Based Simulation for Downtown Revitalization

4.3.1 Simulation Framework

As shown in Fig. 4.1 the whole framework includes three steps, which are dataset preparation, simulation and validation, respectively. Firstly, to test the possibility for visualizing policy impact of downtown revitalization by using the agent-based relocation model, a hypothetical urban space has been prepared for representing planning conditions of a typical monocenter city in Japan. The detail of this virtual urban space has already been introduced in Chap. 2. Then, the prepared dataset will be input into the model for simulation on household residential relocation. During simulation, users can observe the relocation choices of household agents in different age groups and the model can domestically statistic the number of household agents who moved to downtown. The reason for conducting this statistic is to support users

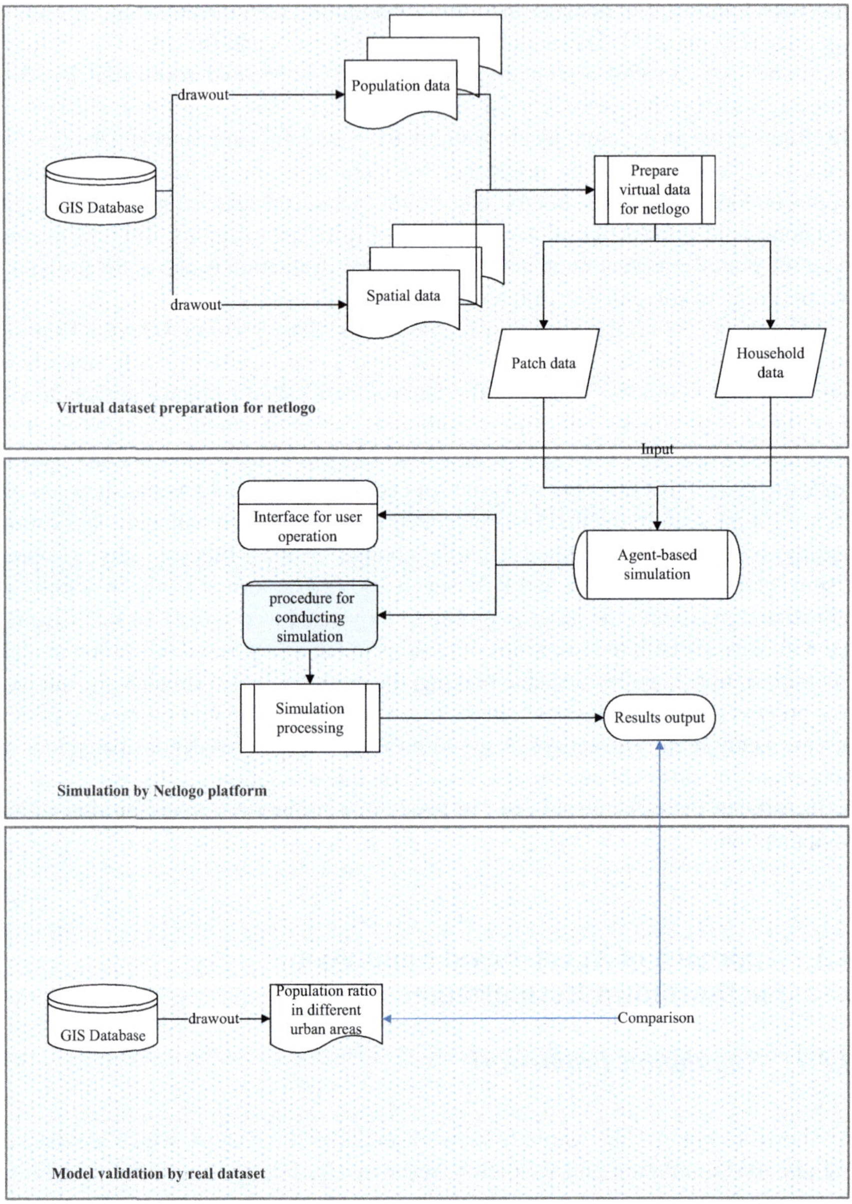

Fig. 4.1 SDL framework for simulation on household residential relocation regarding to downtown revitalization

to analysis whether the promoting policy plays a role in accelerate household remove to downtown. Meanwhile, helping us gain insight on the policy effects on downtown revitalization. Last, the simulation result will be compared with the real data of case study area. This process will be taken as the third step of the framework, namely model validation. It is also an essential part of model development and simulation application.

As shown in Fig. 4.1, the first step of our work is dataset preparation using GIS-based data source. We use it to prepare the virtual space (namely the world or environment for agent behaviors) in order to conduct simulation on the platform of NetLogo. The dataset is mainly related to spatial information including planning information, (such as promotion and control areas of urban planning, land use zonings and land use suitability) household information (such as population density, population ages, household number and their income percentage) in each cell of the hypothetical urban space. These spatial and socio-economical features should be consistent with the planning policy of urban spatial control and socio-economical characteristics. We prepare these spatial attributes of cells as the environment for the simulation of agents' behaviors on residential relocations. These datasets were firstly gained from GIS data source, and then we converted them into raster files by ArcGIS. After the conversion we input them into netlogo to create spatial dataset for our simulation, which is defined as patch data in Fig. 4.1. Meanwhile, according to the relationship between spatial information and census survey of case study area, the household agents have been distributed on the environment, hereafter we named it as virtual urban space. In this work, the number and attributes of household agents are generated according to the household density in each land use zones. The created household data was saved and defined as household data in Fig. 4.1. All the pre-created data will be input into the agent-based simulation to support the decision-making of the strategic plan for downtown revitalization.

4.3.2 Framework of Agent-Based Relocation Model

As we introduced in Sect. 4.2, in this model we will consider both incentive move and adjustment move. For the former we designed a lifecycle change module in the model to simulate one household agent's lifecycles. It is supposed that when one household agent facing to the change of his/her lifecycle stage, he/she shows a big potential in incentive move. While it does not mean household agents will not move if his/her lifecycle stage keeps no changes. We also suppose that if he/she feel unsatisfied with current location, the potential for relocation is high. But the final decision on move or not is simulated by the relocation desire module. If the agent determined to move, a relocation module is designed to simulate this process. Thus, as shown in Fig. 4.2, the relocation model includes three functions: household lifecycle simulation, evaluation on household relocation desire, and simulation of household relocation choice.

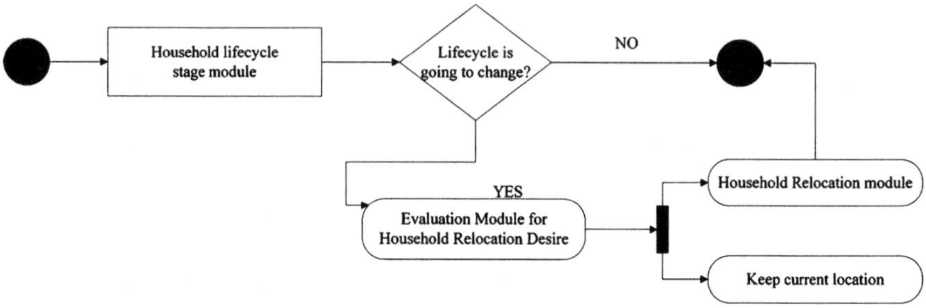

Fig. 4.2 UML state diagram of the agent-based relocation model

4.4 Model Design

4.4.1 The Agents

Within the agent-based relocation model, real persons are not seemed as individual agents but members of a household. A household is an agent, which is a coherent unit in our model. An agent will make decision as a single entity. This single entity is assumed to be composed of one or more people that representing a family. All the household agents are designed to experience a lifecycle process showed in Fig. 4.3. In the model, the household agent will have multiple socio-economic attributes just like what we have in real society. These attributes we considered in the model are household age, members, marital status, incomes, deposits.

As shown in Fig. 4.3, there are seven stages of an agent's lifecycle. The first stage is one agent grow up to 18 years old and leave original household to be an independent household agent. In the model a new agent will be create to represent this process. The created new agent will marry someone after several years' single life. When he/she married, the agent enters the second lifecycle stage "married". In this stage, the agent who are going to marry may show a relocation desire for a new and larger house. Since they married, two single agents become one new married agent in the model. After several years' married life, the coupled housed agent would give birth to babies. For raising children, the married agents may desire for a larger house with good accessibility to school. Thus "raising children" is the third stage of one's lifecycle and the agent in this stage may show a relocation desire again. Till the children of the married household grow up to 18 years old, they will leave their original family and become a new generation of household in our model. For their parents, they enter their fourth lifecycle stage that is named "Children leave house" in Fig. 4.3. In the meantime, the children who left original family will begin their first lifecycle stage, namely "independent". a new generation of households is created, and, as shown by a dotted line, when a child in a household grows up to 18 years old, he/she will find a new residence and begin the first lifecycle stage of a new household. During this process, the individuals who remain in the old household continue to age and eventually retire. One of the individuals will eventually

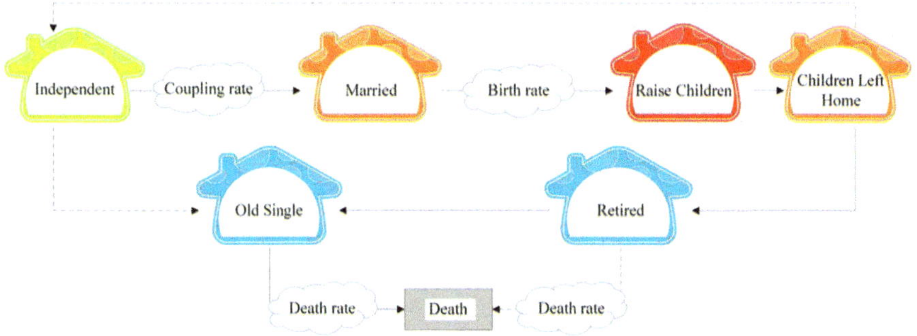

Fig. 4.3 the lifecycle stage of household agent

die, and the remaining individual becomes single again. Finally, the remaining individual will die and disappear from the simulation model.

4.4.2 *Agents' Decision Rule*

1. Whether make a relocation?

For designing the relocation model, we consider there are two kinds of reseaons may lead to households' relocation behaviors, one is resulted from the change of household lifecycle stage and the other one is resulted from unsatisfaction with current location. While lifecycle change may not absolutly lead to relocation, because if one household feels satisfied with current location, he/she does not need to make relocation even lifecycle changed. Thus, we suppose that satisfaction is the direct reseason and also a basis of relocation behavior. It means in our model when one household agent feel unsatisfied with current location he/she will produce a high desire for relocation no matter his/her lifecycle changed or not. Figure 4.4 shows the decision making flow of whether an agent need to make a residential relocation. In the figure, the agent will firstly evaluate his/her satisfaction with curren location. Here we set a sthreshold in the model to recognize the agents who are unsatisfied with current locations. For agent i evaluating his/her satisfaction with current location, the influences of lifecycle stage have been considered that agents in different lifecycle stages have different preferences on residence. When the evaluation result of satisfaction is smaller than the threshold, the agent faces to make a final decision on whether relocating. This process depends on his/her affordability for relocation. It means household i will predict the possible costs of relocation and compare with his/her savings. We suppose that local governemnt may subsidy for households if their relocation choices are downtown. In this case hosueholds will compare the total amount of savings and allowance with possible costs and make a decision on whether move. If the final decision is moving, then relocation process will be simulated in different lifecycle groups. As shown in Fig. 4.4, the lifecycle stage of

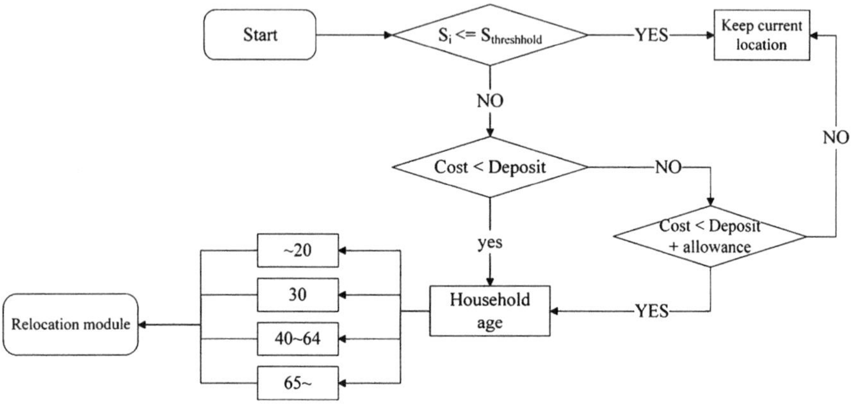

Fig. 4.4 The simulation flow of agents' decision rule for relocation desire

household agents are connected to their ages, which are the groups below 20 years old, 30–39 years old, 40–64 years old and the group elder than 65, respectively.

4.5 How to Make a Relocation?

Residential relocation is widely concerned by researchers in urban study area due to its close and deep relationship with the evolutionary process of urban spatial structure and social-spatial differentiation (Wang et al. 2019; Mao and Wang 2020). As many researches improved and we also introduced above, household's satisfaction with current residence plays a very key role in relocation (Li and Wu 2013; Crull 1979). While the satisfaction may effected by many factors, including the living environment of the residence, building environemnt and facilities around the residence, etc. But nobody can forecast his/her satisfaction with a residence before they living there for a period. So that the utility of residences will be evaluated by households. It is supposed that if a residence provides more utility to a household, he/she would have a high probability on choosing it as a residential location. In this research, in order to investigate the policy effectiveness of promoting households relocation to downtown from other urban areas, we classified the location alternatives for households relocation choices into City Center Area (CCA), Urbanization Promoting Area (UPA) and Urbanization Constraint Area (UCA). Thus, households will follow a decision rule as shown as Fig. 4.5 to make relocation chioces. It means we will only simulate the residential mobility within the above three urban regions. During households agents make relocations, the utility comparision will be conducted between the three regions too. Household agents will choose the region that provide the biggest residential utility to him/her.

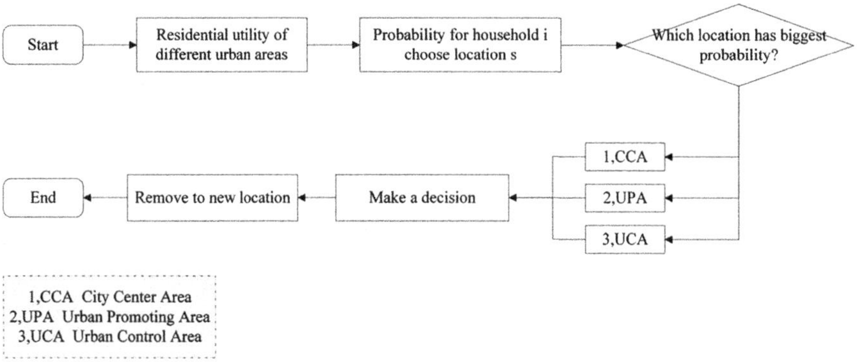

Fig. 4.5 The simulation flow for agents' relocation choices

4.6 Evaluation of Residential Utility and Satisfaction

4.6.1 Equations for Calculating Residential Satisfaction

It is supposed that household's satisfaction with a residential location is produced by building and its surrounding environments. Thus, we defined 18 factors to represent the spatial features of a residential land use cell (as showed as Table 4.1).

In the model, household agents will evaluate their satisfactions with current residential locations through Eqs. 4.1, 4.2, 4.3 and 4.4. In these equations, S_i stands for the satisfaction of household i with his/her current residential location and x_{ij} is the satisfaction of household i produced by variable j. In our work the x_{ij} is divided into four grades. Each grade is assigned by different value which is (1) extremely unsatisfied, (2) unsatisfied, (3) satisfied, and (4) very satisfied, respectively. In the meantime, the b_{uij} in the equation is a vector of retrospective coefficients to variable j. It stands for the attention of household i to spatial variable j when he/she evaluate his/her satisfaction with current location. In this case we need to consider the effects of household lifecycle stage to this coefficient due to different households have different attention on residential utilities while show similarity in the colony who are in the same lifecycle stage. The variables that household agents used to evaluate S_{ij} is showed in Table 4.1.

If one household agent's satisfaction with the current location is less than the threshold, this agent will consider relocating. However, the decision to relocate will be made based on the utility of the alternative locations. The satisfaction of a household with his/her current location can be evaluated based on the attributes of the household agents and the spatial features of the residential locations. The equations for residential satisfaction evaluation are showed below:

Table 4.1 Variables for utility calculation and satisfaction evaluation by agents

X_1	Building size	X_{11}	Building security from earthquake and typhoon
X_2	Building security from fire	X_{12}	Building impairment
X_3	Barrier-free structures for old people	T_1	Comprehensive evaluation on building quality *
X_4	Surrounding safety equipments	X_{13}	Safety while walking on surrounding pavement
X_5	Crime rate	X_{14}	Air or noise pollution
X_6	Accessibility for work or school	X_{15}	Shopping convenience
X_7	Distance from community hospital	X_{16}	Cultural facilities (e.g. distance from library)
X_8	Park or playing ground for children	X_{17}	Green space
X_9	The areas of out space	X_{18}	Street landscape
X_{10}	Communication feasibility with neighbors	T_2	*Comprehensive* evaluation on building's surrounding environment *

$$S_i = \frac{\sum_{j=1}^{n}\left(b_{ugij} \times x_{ij}\right) + \varepsilon_i}{n} \quad \left(\mathrm{n} = 1,\ldots,18\right) \mathrm{and} \left(\mathrm{u} = 1,\ldots,6\right) \tag{4.1}$$

$$0 < S_{threshold} < 4 \tag{4.2}$$

$$S_i > S_{threshold} \quad \left(\text{Satisfied with current location}\right) \tag{4.3}$$

$$S_i \leq S_{threshold} \quad \left(\text{Dissatisfied with current location}\right) \tag{4.4}$$

In which, S_i is household agent i's satisfaction with his/her current residential location, b_{ugij} refers to the residential preferences of household i from income group g in the step of lifecycle u to variable j. The variable x_{ij} used for satisfaction evaluation is shown in Table 4.1.

4.7 Agents Evaluate on the Utilities of Candidate Cells in the Urban Space

It is supposed that household agents will choose the residential locations that provide them with the biggest utilities from the candidate cells. The equations for household agents evaluating residential utility are showed by Eqs. 4.5 through 4.6. In order to reflect the uncertainty of household utility evaluation, a random ε_{is} is set to reflect the unobserved contribution to utility. This random element follows a Gumble distribution and can be generated by Eq. 4.7, in which r follows a random uniform distribution, and constants are set to be 0.5 and 2, respectively.

$$U_{is} = V_{is} + V_{iinter} + \varepsilon_{is} \tag{4.5}$$

$$V_{is} = \sum_{j=1}^{n} b_{ugij} x_{ijs} \quad (n = 1,2,3,...18) \tag{4.6}$$

$$\varepsilon_{is} = \mu - \beta \ln(-\ln r)(\beta > 0),(0 < r < 1) \tag{4.7}$$

$$Q_{is} = \frac{e^{V_{is}}}{\sum_{i=1}^{m} e^{V_{is}}} \tag{4.8}$$

In Eq. 4.6, V_{is} is the residential utility provided by location s. x_{ijs} is the score of location s for variable j (variables are listed in Table 4.1) evaluated by household i. The b_{ugij} is set to reflect the different utility preference of household agents. Q_{is} is the probability of household i choosing location s, which is showed by Eq. 4.8 above. Due to the same residential location may produce different residential utility to different households, we thus designed a decision rule as shown as Fig. 4.6 to represent the different utility preferences of household agents. We classified 6 residential utility preferences of household agents basing on their lifecycle stages (as shown as Table 4.2), in which β_1 is a random perturbation with mean 0 and standard deviation 0.1 generated with a normal distribution.

4.7.1 Agents' Interactions

Although widely used in residential location choices, utility model conventionally does not reflect the influences of individuals' preferences on others. In order to reflect the agents' utility evaluation affected by other agents, we propose that household interactions for residential relocation will happen at global and neighborhood levels. The global influence represents the policy attitude, namely the proportion of households in a city that accept and plan to use the local residence promotion policy for relocation. At the neighborhood level, neighborhood influence will be considered in order to represent the effect of neighbors on household residential relocation. It reflects the sharing of information about the allowance policy by households within a small neighborhood community. The details of these two factors are explained below:

1. Neighborhood influence (x_{i19}): ratio of neighbors that use the residential allowance policy to relocate to a downtown area N_{move} divided by the number of neighbors that do not use the policy for relocation N_{nomove}. Here, x_{i19} can represent agent choices influenced by neighbors who plan to use the policy. This indicator is added to the utility model as an extra component of utility for relocation.

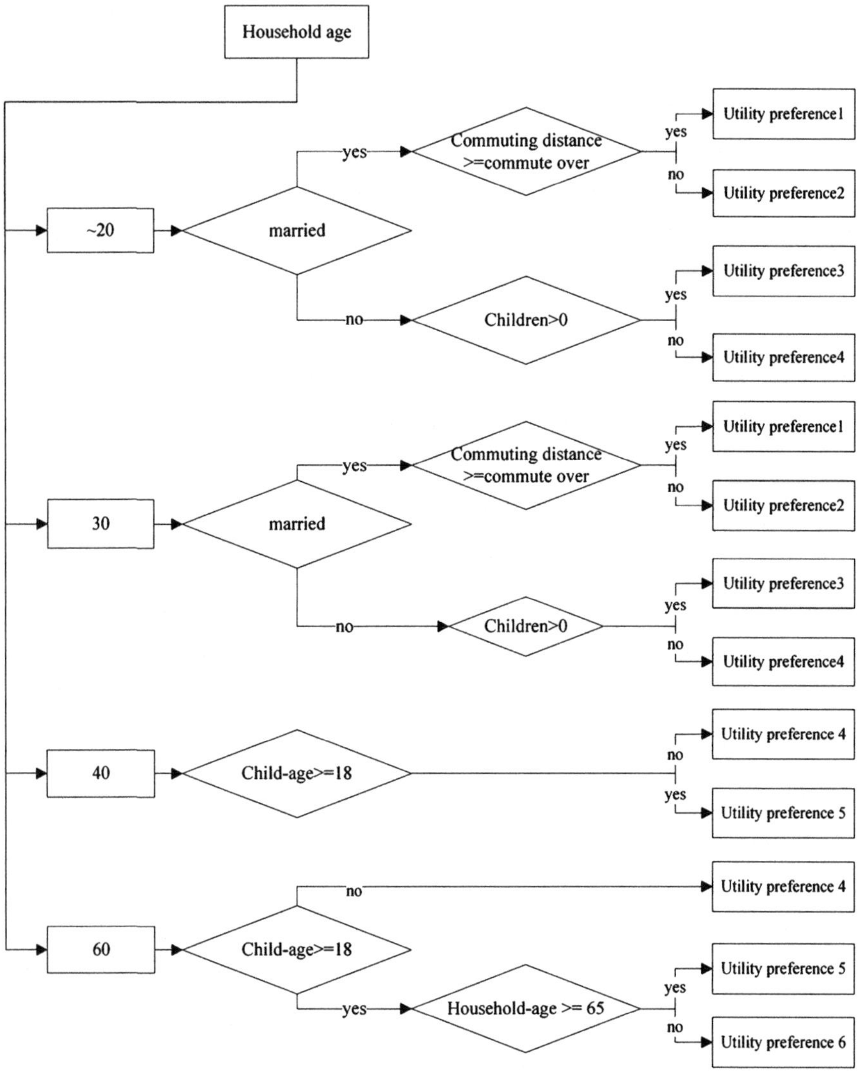

Fig. 4.6 Utility preferences of household agents for residence

2. Global influence (x_{i20}): ratio of the total number of households that use the policy for relocation G_{move} divided by the total number of households G_{total}. This factor represents the proportion of households in the city that accept the policy and plan to use the policy for residential relocation to a downtown area. This factor is defined as a component of utility, which is the policy impact on individual relocation decisions for the entire city.

Table 4.2 Parameters for utility preference by households in different lifecycle stages

Utility preference 1	Single households would like living far away from work place	$\{b_{ugij}\} = \{b_{1gj}(1+\beta_i)\mid j = 1,2,...18\}$ $b_{1gj} = b_j + a + \alpha_{1gj}$ where, if j = 3,6,7,13; a = −0.01. Otherwise a = 0 (g stands for poor, middle and rich)	Poor: if j = 1,9,17,18 $\alpha_{1gj} = -0.01$ Middle: if j = 1,5,15 $\alpha_{1gj} = 0.005$ Rich: if j = 1,5,9,17,18 $\alpha_{1gj} = 0.01$
Utility preference 2	Single households would like living near to work place	$\{b_{ugij}\} = \{b_{2gj}(1+\beta_i)\mid j = 1,2,...18\}$ $b_{2gj} = b_j + a + \alpha_{2gj}$ where, if j = 3,7,13; a = −0.01. Otherwise a = 0 (g stands for poor, middle and rich)	Poor: If j = 1,9,17,18 $\alpha_{2gj} = -0.01$ Middle: If j = 1,5,6 $\alpha_{2gj} = 0.005$ Rich: If j = 1,5,6,9,17,18 $\alpha_{2gj} = 0.01$
Utility preference 3	Married households without children	$\{b_{ugij}\} = \{b_{3gj}(1+\beta_i)\mid j = 1,2,...18\}$ where, if j = 3,7,8,13; a = −0.01. Otherwise a = 0 (g stands for poor, middle and rich)	Poor: If j = 9,17,18 $\alpha_{3gj} = -0.01$ Middle: If j = 1,5,6,15$\alpha_{3gj} = 0.005$ Rich: If j = 1,5,9,17,18 $\alpha_{3gj} = 0.01$
Utility preference 4	Married households with children	$\{b_{ugij}\} = \{b_{4gij}(1+\beta_i)\mid j = 1,2,...18\}$ where, if j = 3,7; a = −0.01. Otherwise a = 0 (g stands for poor, middle and rich)	Poor: If j = 9,17,18 $\alpha_{4gj} = -0.01$ Middle: If j = 1, 5, 6, 8, 9,15,16 $\alpha_{4gj} = 0.005$ Rich: If j = 1,5,8,9,17,18 $\alpha_{4gj} = 0.01$
Utility preference 5	Children elder than 18 years old	$\{b_{ugij}\} = \{b_{5gj}(1+\beta_i)\mid j = 1,2,...18\}$ where, if j = 8; a = −0.01. Otherwise a = 0 (g stands for poor, middle and rich)	Poor: If j = 1,9,17,18 $\alpha_{5gj} = -0.01$ Middle: If j = 5,14,15 $\alpha_{5gj} = 0.005$ Rich:If j = 5,9,14,157,18 $\alpha_{5gj} = 0.01$
Utility preference 6	Retired households	$\{b_{ugij}\} = \{b_{5gj}(1+\beta_i)\mid j = 1,2,...18\}$ where, if j = 6,8,16; a = −0.01. Otherwise a = 0 (g stands for poor, middle and rich)	Poor: If j = 3,7 $\alpha_{6gj} = 0.001$ Middle: If j = 3, 5, 7, 9, 14,15 $\alpha_{6gj} = 0.005$ Rich: If j = 3, 5, 9, 10, 14,17,18 $\alpha_{6gj} = 0.01$

4.8 Environment and Parameters for Simulation Configuration

4.8.1 Hypothetical Urban Space and Household Agents

Regarding to the creation of environment as a virtual urban space for accommodating household agents has been introduced in Chap. 2 very detail, we will not repeat here. In this work we directly use the created environment and distributed agent data for further simulation. The virtual urban space and household distribution is showed by Fig. 4.7. In the following simulation each simulation loop stands for 5 years.

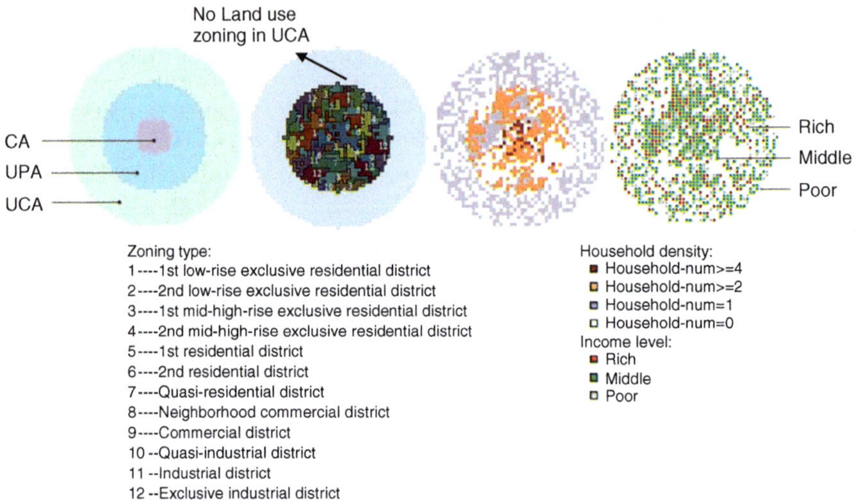

Fig. 4.7 The virtual urban space for accommodating household agents

4.8.2 Parameters for Satisfaction Evaluation and Utility

4.8.2.1 Questionnaire of Residential and Housing Environment in Kanazawa City

As we introduced in Chap. 1, agent decision rules are basing on existing researches, laboratory observation or analyzing on the big data of individual behaviors. In this work we use the utility model to simulate agent behaviors in residential relocation choices, which has already been proved by literatures. While how will an agent evaluate on a cell's residential utility? Existing literatures already revealed the factors one individual will evaluate when making a residential location. But in practice we always find a phenomenon that the same residence makes somebody satisfied but others maybe unsatisfied. It means although the same spatial features, the residence will produce different utilities to different individuals. This is because each household has different residential utility preference. In this study, we need to test and recognize these varied preferences for simulation. It could be done by many means, such as questionnaire, long term laboratory observation or big data analysis, ect. Due to the big data of household residential relocation in case study area is not available, we finally chose questionnaire to investigate households residential statues and relocation desire. We thereby estimate the coefficients to the factors in utility model through questionnaire.

During December 2009 to January 2010, we conducted a questionnaire investigation of residential environment and housing environment in Kanazawa City, Japan. The brief introduction of this investigation is showed by Table 4.3. As showed by this table, the investigation took the main suburban areas of Kanazawa City as investigation target areas. The respondents from the six suburban areas were respectively 196 in Nukashin Machi, 202 in Mgae Machi, 133 in Minma Machi and 209 in

Table 4.3 Investigation of residential environment and housing environment in Suburban area of Kanazawa City

Investigation time	Investigation areas	Responders	Collecting rate (%)
Dec 2009–Jan 2010	Nukashin Machi	124	63.27
	Magae Machi	139	68.81
	Minma Machi	90	67.67
	Kubo Machi	130	62.20
	Izumino Machi	125	64.10
	Izumigaoka Machi	134	62.33

Table 4.4 Investigation contents of residential environment and housing environment in Suburban area of Kanazawa City

Topics	Questions
Topic 1: Current residence and its residential environment	1, how do you get current residence? 2, when have you been to current residential location? 3, where did you live before this residence? 4, why do you choose current area for residential location? 5, why do you choose current residence or land? 6, satisfaction evaluation on current residential environment and housing environment 7, what do you feel about current location area?
Topic 2: Future demand for residential location	1, would you like to move to a new residential location? 2, if choose "1 or 2" within answer list of question 1, please answer why 3, if choose "3 or 4" within answer list of question 1, please answer why 4, if you plan to do residential relocation, what kind of house or land do you want? 5, do you have inheritor for your current residence 6, how do you think about the housing and residential environment in the future?
Topic 3: Residential policy in Kanazawa City, Japan	1, do you know the residential promoting policy in Kanazawa city? 2, do you plan to move to downtown by using this policy and receiving an allowance for your relocation? 3, would you move to downtown for a residential relocation by using this policy?
Topic 4: Neighborhood	1, what is your ideal neighborhood? 2, what is the real neighbourhood of your current residence? 3, do you join community activities often? 4, do you think that the community activities are helpful for your daily life?
Topic 5: Commuting means	By assuming all housheolds have car in this work, we thereby havenot listed the detailed questions of this topic

Kubo Machi, 195 in Izumino Mchi, 215 in Izumigaoka Machi. The contents and the detailed questions for all topics are introduced in Table 4.4. In this work we concentrate on the topics through topic 1–3. We assume that all the households have cars in Suburban areas so we have not listed the detailed questions regarding to topic 5. In 4.4.2.2 we will estimate the parameters for conducting household satisfaction and utility evaluations in the modeling basing on the investigation results regarding to this questionnaire.

4.8.2.2 Parameters for Household satisfaction Evaluation and Utility Calculation

Regarding to the first topic of current residence and its residential environment, households were asked to evaluate on the 18 factors we listed in Table 4.5 to represent their satisfaction with their current residence. These variables are the same with what we listed in Table 4.1 except for the two comprehensive factors. The question 6 of topic 1 in Table 4.4 divided the satisfaction degree of each evaluation variables into four levels and households were asked to discriminate score from these four levels for each of the variables in Table 4.5. As shown in Table 4.4, a concluded question has been asked as "would you like to move to a new location?" by question 1 in topic 2. The results were quantified by −1, −2, 1 and 2 four levels, respectively stands for not move, not move if I can, move if I can and move. According to the investigation results we estimate the parameters through regression analysis between S_i and x_{ij} by statistic software R to recognize the coefficients b_j (b1 to b18).

Furthermore, policy interaction for household satisfaction evaluation in household residential relocation desire module of MODEL has been taken into account. As shown in Table 4.4 of the questionnaire, the topic 3 focused on the residential promoting policy in Kanazawa city. The detailed options of this question are showed in Table 4.6, within which there are six choices. Basing on the response of this question we try to figure out how will questions in Topic 1 impact on households' interests in residential relocation. To achieve this purpose, we add two factors in the utility model, namely x_{i19} and x_{i20}. Then we estimated the regression correlation of S_i and x_{ijs} to determine the coefficients b'_j *(b_1 to b_{20})* for reflecting the policy effects on household relocation choices. The estimation results are listed in Table 4.5 as b'j.

4.8.2.3 Initial Parameters for Simulation

Besides the parameters for agents evaluating on residential utilities and satisfactions, when the simulation run usually initial parameters are necessary. These parameters could be divided into global parameters that are related to global variables of the simulation model, spatial attributes that are related to the environment of the simulation model and agents' attributes that are related to agents. Table 4.7 shows us the parameters and spatial features we used in modeling. In the table we set the global parameters for household lifecycle stage basing on the Census Survey of case study area (in this chapter, Kanazawa city Japan is chose as the case study area).

Table 4.5 Partial correlation coefficient of variables on household relocation utility and satisfaction

Variables for satisfaction evaluation	Coefficients (bj)	Partial coefficient (no policy interaction)	Significant	Coefficients(b'j)	Partial coefficient (with policy interaction)	Significant
x1: Building size	b_1	0.170634	****	b'_1	0.069904	.
x2: Building security from earthquake, typhoon	b_2	0.140753	****	b'_2	0.153358	***
x3: Building security from fire	b_3	0.045609	.	b'_3	0.085280	*
x4: Building impairment	b_4	0.029561	.	b'_4	0.112268	**
x5: Barrier-free structures for old people	b_5	0.067244	.	b'_5	0.051591	.
x6: Surrounding safety equipments	b_6	0.125595	***	b'_6	0.072290	*
x7: Safety while walking on surrounding pavement	b_7	0.126829	***	b'_7	0.105173	*
x8: Crime rate	b_8	0.109048	***	b'_8	0.085835	*
x9: Air or noise pollution	b_9	0.162371	****	b'_9	0.098698	*
x10: Accessibility to work or school	b_{10}	0.056896	.	b'_{10}	0.079705	*
x11: Shopping convenience	b_{11}	0.118102	***	b'_{11}	0.075495	*
x12: Accessibility to community hospital	b_{12}	0.111277	***	b'_{12}	0.052196	.
x13: Cultural facilities (e.g. distance from library)	b_{13}	0.097474	**	b'_{13}	0.058622	.
x14: Park or playing ground for children	b_{14}	0.100275	**	b'_{14}	0.100814	*
x15: Green space	b_{15}	0.079093	**	b'_{15}	0.123672	**
x16: The areas of out space	b_{16}	0.172469	****	b'_{16}	0.027558	.
x17: Street landscape	b_{17}	0.087435	**	b'_{17}	0.102831	*
x18: Communication feasibility with neighbors	b_{18}	0.061328	.	b'_{18}	0.020922	.
Interaction 1: Neighborhood influence (x_{i19})				b'_{19}	0.064556	.
Interaction 2: Global influence (x_{i20})				b'_{20}	0.174542	****

Significant values: 0.1 "." 0.05 "*" 0.001 "**" 0.0005 "***" 0.00001 "****"

Table 4.6 Question regarding to residential promoting policy

Question 2 of topic 3	Alternatives
Do you plan to move to downtown by using this policy and receiving an allowance for your relocation?	1.Plan to buy house in downtown by using the policy 2. Plan to buy apartment by using the policy 3. Planned to buy house by using the policy 4. Planned to buy apartment by using the policy 5. Won't use the policy 6. Others

Table 4.7 Initial parameters and spatial features for household residential relocation modeling

	Parameters	Predefined data
Global parameters	Birth rate	0.90%
	Death rate	0.80%
	Coupling rate	0.59%
	Threshold for satisfaction	0.1
	Scenario option	With residential promotion policy or without this policy
	Parameters for setting random numbers	For random numbers in various distributions in simulation model
	Commute-over	Maximum commuting distance
Attributes of virtual space	Land use zoning	12 types of land use zonings
	Household density	Household numbers based on floor area ratios based on land use zoning
	Planning areas	CCA(4%), UPZ(32%), UCA(64%)
	Other attributes of	18 variables (as Table 4.5) evaluated based on location.
	House price	800 (CCA), 400(UPA), 200(UCA) thousands (JPY)/ 3.3 m^2
Attributes of household	Income	Low(20%), middle (60%) and high (20%)
	Car ship	Yes
	Household age	The 6 groups described in Fig. 4.2
	Family members	1–6 based on lifecycle stages
	Saving	Random number based on income groups and lifecycle stages
	Global influence	As described in Sect. 2.5
	Neighborhood influence	As described in Sect. 2.5
	Personal Satisfaction threshold	The global threshold 0.1 and a random number in normal distribution with mean 0 and SD 0.01.
	Satisfaction of current location	As described in Sect. 2.4.1
	Individual parameters of utilities regarding relocation alternatives	As described in Table 4.2.
	Location	Coordination x, y

4.9 Model Test

4.9.1 Sensitivity Test of Household Lifecycle

Because the utility model has already widely accepted and utilized by researchers, so there is no need to verify it in our work. Instead of, we conduct the sensitivity analysis regarding to lifecycle stage and policy interactions. The interface of the simulation model is showed as Fig. 4.8. In order to employ the agent-based model for decision-making support of planning policy, the parameters and variable of the model must follow the real planning conditions. Thus, the virtual urban space is simplified and designed regarding to the same conditions in Kanazawa City, a typical Japanese local city. For simplify the simulation model, we propose that commercial center in urban center and the job opportunities are all centralized in CA areas.

The model sensitivity test of birth rate influences on household number is showed by Fig. 4.9. Within this figure we simulated three different birth rate, which are 0.1, 0.5 and 1.0 respectively. In the meantime, we keep death rate and coupling rate in stable with their initial values. The model behaviors are divided into two parts, the first part is about household numbers in different urban areas that shown in the top of Fig. 4.9. As shown by these three pictures, the household number in UCA area are obviously more than CA areas, while when the birth rate grow to 1.0, the household number in UPA decreased less than CA area, while during the whole process the household number in UCA is majority than the number in CA area. As shown in the bottom of this figure, the household numbers are simulated by four household age levels. The growth of birth rate leads to the increase of young households (younger than 20 years old), which slowed down the situation of population aging. From this viewpoint, the increase of birth rate would relieve the population aging but cannot increase the residential rate in urban center area. Thus, without policy intervention although the household number is increasing, downtown decline cannot be relieved.

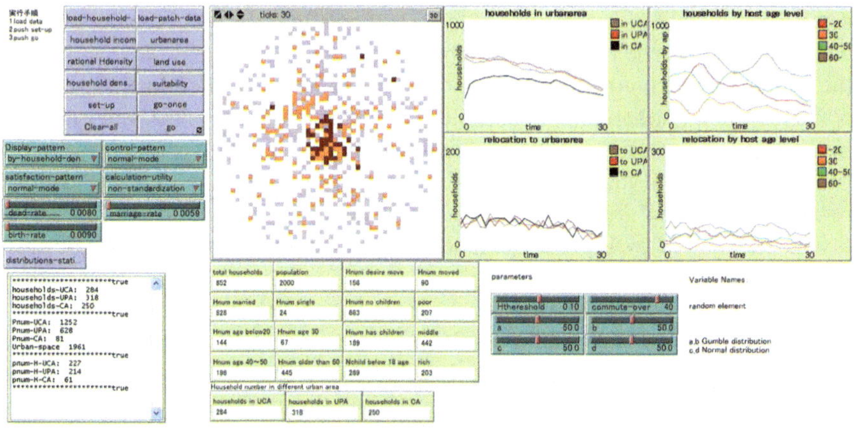

Fig. 4.8 The interface of MODEL in Netlogo platform

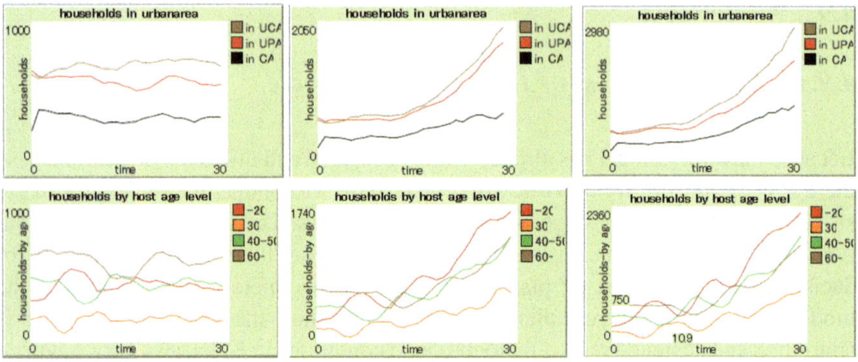

birth rate = 0.1 birth rate = 0.5 birth rate = 1

Fig. 4.9 Household number change according to birth rate regulation

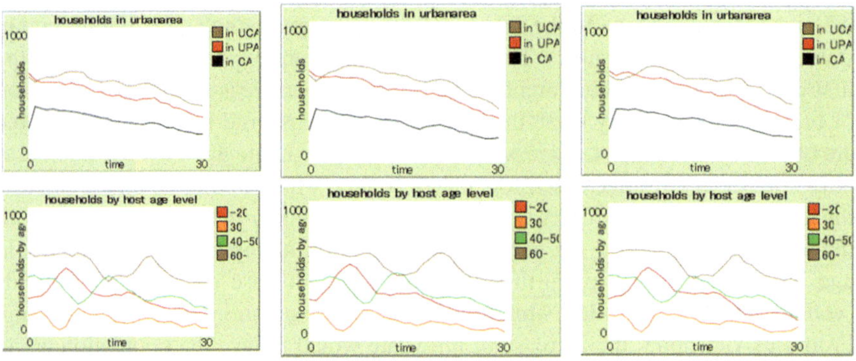

death rate = 0.1 death rate = 0.5 death rate = 1

Fig. 4.10 Household number change according to death rate regulation

Contrarily speaking, if we keep the birth rate stable but increase the value of death rate, obviously as shown as Fig. 4.10, not only the total number of households decreases but also the household number in different urban areas. We can see that the bottom picture on the right of Fig. 4.10, the households elder than 60 years old is the majority. Furthermore, households around 30 years old is the minority. This result helps us get a sight in the situation that the increase of birth rate will play a positive role in relieving the population aging but would not useful in downtown revitalization. Even the death rate decrease, the situation is also the same. The results are showed by Fig. 4.11, which are quite similar with the results of death rate change.

Anyway, model tests regarding to the three factors relating to household lifecycle, the results show that the key influences are coming from birth rate change. While the increasing the birth rate only can increase the number of young households and the frequency relocations but most relocations still occur in the UCA. This is far away from the policy target that accelerate households relocate to CA area. As

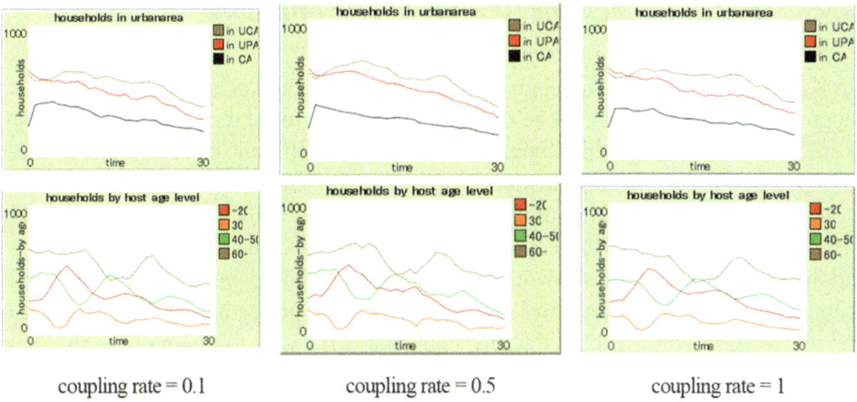

Fig. 4.11 Household number change according to coupling rate regulation

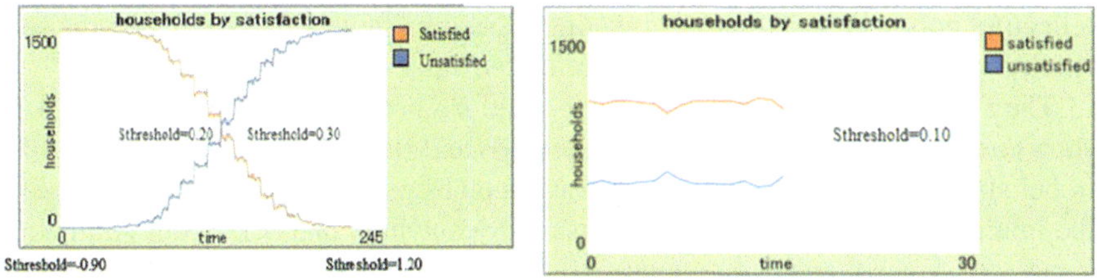

v. 4.12 Satisfaction threshold influence on household residential relocation desire

a result, we can conclude that in the modeling, the changes in the parameters of the lifecycle stage can produce reasonable changes in household residential relocation and the changes are sensitive too.

4.9.2 Satisfaction Threshold for Making Decision on Relocation Choices

In Fig. 4.12 we test the sensitivity of satisfaction threshold for evaluation on household residential locations. As shown on the left of Fig. 4.12, the number of households satisfied by current location is decreasing by the increase of threshold. At the cross point of threshold equate to 0.2 and threshold equate to 0.3 stands for the state that the proportion of households who satisfied with current locations takes 50% of total households in local city. While existed researches show that in Kanazawa city the proportion of households who unsatisfied with current location and desire to relocation takes up 30% (Kikuchi and Nojima 2007; Kawakami and Takama 1978). Through many times of experiments, we found out that when we set the satisfaction threshold as 0.10, the simulated proportion of household agents who desire for relocation take up 30% of the whole agents (as shown as the right picture of Fig. 4.12). Finally, the threshold in the simulation model is set as 0.10.

4.9.3 Policy Impacts on Downtown Revitalization

In Kanazawa city, Japan, local government pays more and more attentions to policy approaches in order to revitalize their downtown and make their city compact. This kind policies have affluent contents and stand for different levels. Some of these policies have been making downtown regeneration efforts including development controls on large-scale shopping centers (Shen et al. 2011), which will be introduced in new chapter. With respect to residential policy, Japanese government pays attention to revitalize downtown through increasing the population density in city center areas. Thus, in Kanazawa city, one residential policy has been carried out to revitalize its downtown by attracting households relocate to downtown. The main strategy of this policy is supplying the residents who moved to downtown with local housing allowances. More details of this residential promoting policy (issued in 2005) are showed by Table 4.8.

In this part we will focus on testing the model through adjusting the values for reflecting policy impacts. As what we introduced in Table 4.5, we will test the sensitivity of policy impacts through adjusting the values of b_{19} and b_{20}.

The simulation results regarding to b_{19} and b_{20} both equate to 0.1, the Fig. 4.13 shows us the spatial distribution of households and related statistic results. Because in our simulation we treat the urban space as a closed space, so we do not consider the immigration or migration from outside. Meanwhile, one tick in our simulation stands for 5 years and we simulate 30 ticks, so the simulation results represent 150 years later from the initial stage. These two conditions take place together and hereby, result in the decreasing of total household numbers during simulation.

As shown by the first picture from the left of Fig. 4.13, the residential density in CA area is obviously higher than other urban areas. In the meantime, the household density in UCA and UPA decreased a lot. Besides the influences resulted from global household number decrease, policy affects also influenced on household residential relocations within the three urban areas. From Figs. 4.13, 4.14 and 4.15, as the values for b_{19} and b_{20} increased, the household number in CA area increased correspondingly. Bigger b_{19} and b_{20} are, more households moved to CA area. While because during this process we kept birth rate, coupling rate and death rate stable, so as showed by Figs. 4.13, 4.14, and 4.15, the proportion of households in different

Table 4.8 The strategy of downtown revitalization in Kanazawa city, Japan

Building types	Building choice		Allowances
House	Buy new house	Single household	10% payment, 2 million JPY
		Two households	10% payment, less than 3million JPY
	Buy or repair old house		Basic part+ supplementary part, less than 500,000 + 200,000 JPY
Apartment	Buy new apartment		Basic part (5% of payment) + supplementary part (1%), less than 1 million +200,000 JPY
	Old apartment	Buy	
		Repair	50% of design payment, less than 1 million JPY

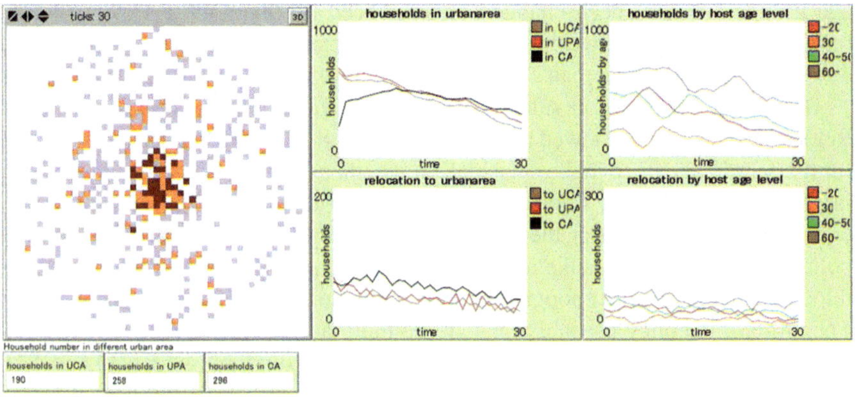

Fig. 4.13 $b_{19} = b_{20} = 0.1$

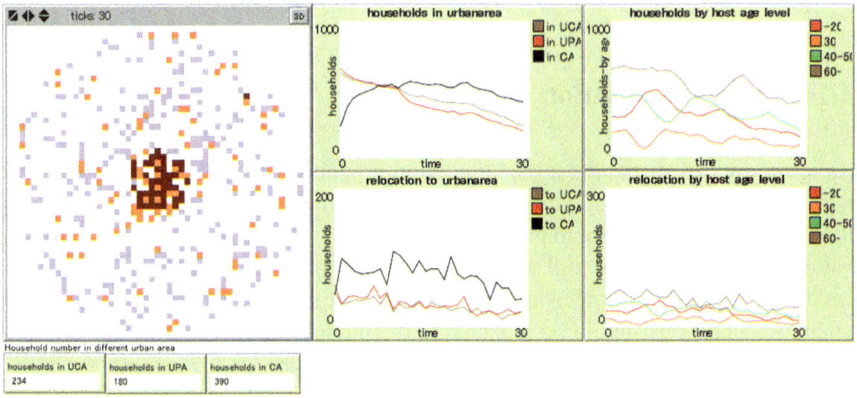

Fig. 4.14 $b_{19} = b_{20} = 0.5$

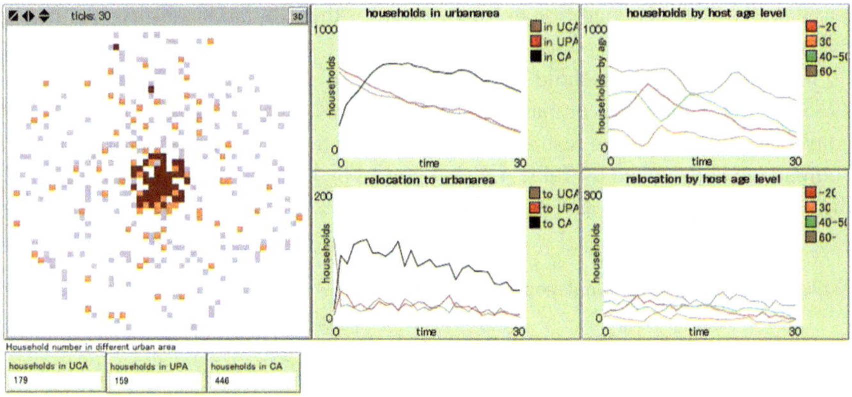

Fig. 4.15 $b_{19} = b_{20} = 1$

age groups have not been changed or influenced by this policy. Meanwhile, the ranks of relocations between different age groups are elderly than 60 years old, 40–50 years old, younger than 20 (around 18) years old and 30 years old, respectively.

Basing on the decision rule we set for the evaluation module of Household Relocation Desire that showed by Fig. 4.4 in Sect. 4.4.2, we defined a necessary step for household agents to make decisions on whether do relocations. This step is asking them to compare the relocation cost with their deposits. If the deposits are not enough, they will consider the allowance supplied by local residential promoting policy. Thus, we could make a conclusion that maybe the limited allowances for relocation tend to be not enough to attract young households relocate to downtown. Therefore, this policy probably tends to be more attractive to the households who are price well-off. Just as shown by the three figures from Figs. 4.13, 4.14 and 4.15, the households elderly than 60 years old, 40–50 years old both showed higher potential in relocation to downtown by using this policy than households who are around 30 years old or younger.

4.10 Model Validation

As tested by the virtual data, the simulation model works well and can reflect the number changes of households in different urban areas during sensitivity analysis of lifecycle stage and household interactions. While to validate the model a real dataset validation of the model is still necessary. As what we have introduced in Chap. 2, if we develop an ABM for decision-making support of a targeted planning policy, the planning conditions should be reflected in the model. in this case, the environment of the ABM is developed on Japan City Planning Law, as a Japanese local city, Kanazawa city also follows the law. In this case, although the urban space for accomodating household agents is virtual, the law for spatial features, household distributions are the same with case study area. Especially we set the parameters for simulation as the real condition in Kanazawa city, the simulation results should have the similar characteristics with case study area.

We simulate the household relocation between different urban area from year 1985 to 2000. Then we compare the simulation results with local statistic data by converting the simulation results into household ratios in different urban areas. The comparison are showed by Table 4.9. As showed by local consensus survey in 1985, the household ratios in CA is 33.9% and 66.1% in UPA and UCA, while the

Table 4.9 Comparison of simulation results and real status

	Years	1985 (%)	1990 (%)	1995 (%)	2000 (%)
Real dataset	CA	33.9	31.9	29.0	26.6
	UPA + UCA	66.1	68.1	71.0	73.4
Simulated results	CA	32.7	32.1	30.4	33.7
	UPA + UCA	67.3	67.9	69.6	66.3

simulation results are respectively 32.7% and 67.3%. The simulation results are quite near to the real data. In 1990, the household ratio in CA is 31.9% in real dataset and 32.1% in simulation results, which are still very near, and so does the situation in 1995. While in 2000 the simulation results of households in CA is nearly 10% higher than real situation. This result would probably belong to the adjustment of the Building Standards Act law in 1998, by which the traditional 11 types of land uses for urbanization control changed to be local freely control (Building Standards Act 1998).

4.11 Discussions

This case shows an example on how an ABM is developed and used in decision-making support. As introduced in this Chapter, before model application an developed model should firstly be tested to prove its validation and parameter sensitivity. Different from conventional residential relocation model developed on utility theory, in this work the agents' behaviors are supposed to be effected by the argued planning policy. It is necessary for policy analysis and decision-making support by using ABMs. It means when we develop an ABM for policy analysis or decision-making support, the interactions between agents need to be linked to the policy, such as policy information sharing or policy attitude sharing. This policy sharing behavior will further effects on one agent's behaviors, for example one agent is not very interested in the residential promotion policy but his/her neighbors are quite interested in it and plan to use it, these kind possitive feedbacks will effects on the agent and let he/she re-consider the utilizetion of the policy, and vice verse.

References

Berger T., Schreinemachers P., Woelcke J., 2006. Multi-agent simulation for the targeting of development policies in less-favored areas. Agricultural Systems, 88(1), pp. 28–43.

Benenson, I., 1998. Multi-agent simulations of residential dynamics in the city. Computers, Environment and Urban Systems, vol. 22, pp. 25–42.

Building Standards Act., 1998. Ministry of Land, Infrastructure, Transport and Tourism, Japan. Japanese website Archived at: http://www.mlit.go.jp/

Babak Mahdavi Ardestani, David O'Sullivan, Peter Davis, 2018. A multi-scaled agent-based model of residential segregation applied to a real metropolitan area, Computers, Environment and Urban Systems, vol. 69, pp. 1–16.

Bruch, E. E., 2014. How population structure shapes neighborhood segregation. American Journal of Sociology, 119(5), pp. 1221–1278.

Crull S. R., 1979. Residential satisfaction, propensity to move and residential mobility: A causal model.

Cao, X., Mokhtarian, P.L., Handy, S.L., 2009. The relationship between the built environment and non-work travel: a case study of Northern California. Transportation Research Part A: Policy Practice, 43, pp. 548–559.

Chen C., Gong H.M., Paaswell R., 2008. Role of the built environment on mode choice decisions: additional evidence on the impact of density. Transportation, 35, pp. 285–299.
Dawn C. P. and Filatova T., 2008. A conceptual design for a bilateral agent-based land market with heterogeneous economic agents. Computers, Environment and Urban Systems, 32, pp. 454–463.
Diappi L. and Bolchi P., 2006. Gentrification Waves in the Inner-City of Milan - A multi agent / cellular automata model based on Smith's Rent Gap theory, Innovations in Design & Decision Support Systems in Architecture and Urban Planning, pp. 187–201.
Ettema D., 2011. A multi-agent model of urban processes: Modelling relocation processes and price setting in housing markets. Computers, Environment and Urban Systems, 35, pp. 1–11.
Fontaine C.M., Rounsevell M.D.A., 2009. An agent-based approach to model future residential pressure on a regional landscape. Landscape Ecological, 24, pp. 1237–1254.
Fatmi M. R., Habib M. A., 2018. Microsimulation of life-stage transitions and residential location transitions within a life-oriented integrated urban modeling system, Computers, Environment and Urban Systems, vol., 69, pp. 87–103.
Fossett, M., 2015. Generative models of segregation: Investigating model-generated patterns of residential segregation by ethnicity and socioeconomic status. The Journal of Mathematical Sociology, 35(1–3), pp. 114–145.
Gaube, V., Remesch, A., 2013. Impact of urban planning on household's residential decisions: An agent-based simulation model for Vienna. Environmental Modelling & Software, 45, pp. 92–103.
Gerber P., Ma T.-Y., Klein O., Schiebel J., Samuel C.-P., 2017. Cross-border residential mobility, quality of life and modal shift: A Luxembourg case study. Transportation Research Part A: Policy Practice, 104, pp. 238–254.
Haase D., Lautenbach S., Seppelt R., 2010. Modeling and simulating residential mobility in a shrinking city using an agent-based approach. Environmental Modelling & Software, 4, pp. 1225–1240.
Jordana R., Birkin M., Evans A., 2014. An agent-based model of residential mobility Assessing the impacts of urban regeneration policy in the EASEL district. Computers, Environment and Urban Systems, 48, pp. 49–63.
Kawakami M., Takama C., 1978. Investigation of household residential relocation desire in Kanazawa City (in Japanese). Architecture Institute of Japan, 13, pp. 67–72.
Kikuchi Y., Nojima S., 2007. Resident's Mind about Residence Selection in Suburban Housing Estates: Case study of 4 suburban estates in Fukui City. City planning review: Special issue on city planning, 42(3), pp. 217–222.
Klaassen, L. and Paelinck J., 1979. The feature of the large towns. Environment and Panning A, 11, pp. 1095–1104.
Klinger T., 2017. Moving from monomodality to multimodality? Changes in mode choice of new residents. Transportation Research Part A: Policy Practice, 104, pp. 221–237.
Lee Q., Park B., Vlek S. J. P.L.G., 2010. Land Use Dynamic Simulator (LUDAS): A multi-agent system model for simulating spatio-temporal dynamics of coupled human–landscape system: 2. Scenario-based application for impact assessment of land-use policies. Ecological Informatics, 5(3), pp. 203–221.
López-Ospina H. A., Martínezb F. J., Cortésb C. E., 2016. Microeconomicmodelofresidentiallocationincorporatinglifecycleand social expectations. Computers, Environment and Urban Systems, 55, pp. 33–43.
Li Li, Xiaojian Li, Beibei Hai, Xuefeng Wang, Jiawei Xu, 2020. Evolution of rural settlement in an inland nonmetropolitan region of China at a time of rapid urbanisation: The case of Gongyi, Journal of Rural Studies, Volume 79, pp. 45–56.
Li Z., Wu F., 2013. Residential Satisfaction in China's Informal Settlements: A Case Study of Beijing, Shanghai, and Guangzhou. Urban Geography, 34(7), pp. 924–949.
Mao, Z., Wang, D., 2020. Residential relocation and life satisfaction change: Is there a di?erence between household couples? Cities. 97: 102565.

Molin, E., Timmermans, H., 2003. Accessibility considerations in residential choice decisions: accumulated evidence from the Benelux. TRB 2003 Annual Meeting CD-ROM, Washington, DC.

Miller, E. J., Hunt, J. D., Abraham, J. E., & Salvini, P. A., 2004. Microsimulating urban systems. Computers, Environment and Urban Systems, 28, pp. 9–44.

Mendonça R., et al., 2020. Assessing economic instruments to steer urban residential sprawl, using a hedonic pricing simulation modelling approach, Land Use Policy, Volume 92, 2020, 104458.

Ma Y., Shen Z.J., Kawakami M., 2013. Agent-Based Simulation of Residential Promoting Policy Effects on Downtown Revitalization. the Journal of Artificial Societies and Social Simulation, 16(2) 2.

MA Y., SHEN Z. J., Nguyen D. T., 2016. Agent-Based Simulation to Inform Planning Strategies for Welfare Facilities for the Elderly: Day Care Center Development in a Japanese City. Journal of Artificial Societies and Social Simulation, 19(4)5.

Moreno A. T., Moeckel R., 2018. Population Synthesis Handling Three Geographical Resolutions. International Journal of Geo-Information, 7, pp. 174.

Noth, M., Borning, A., Waddell, P., 2003. An extensible, modular architecture for simulating urban development, transportation, and environmental impacts. Computers, Environment and Urban Systems, 27(2), pp. 181–203.

Rieniets T., 2009. Shrinking cities: causes and effects of urban population losses in the twentieth century. Nature and Culture, 4(3), pp. 231–254.

Shen Z. J., Yao A. Xiaobai, Kawakami M., Chen P., 2011. Simulating Spatial Market Share Patterns for Impacts Analysis of Large-Scale Shopping Centers on Downtown Revitalization, Environment and planning B: planning and design, vol. 38(1): 142–162.

Tannier C., Curab R., Leturcq S., Zadora-Rioc E., 2020. An agent-based model for exploring the combined effects of social and demographic changes on the concentration and hierarchy of rural settlement patterns in North-Western Europe during the Middle Ages (800–1200CE). Journal of Anthropological Archaeology, 59, pp. 101204.

Waddell, P., Borning, A., Noth, M., Freier, N., Becke, M., & Ulfarsson, G., 2003. Microsimulation of urban development and location choices: Design and implementation of UrbanSim. Networks and Spatial Economics, 3, pp. 43–67.

Wang F. L., Mao Z. D., Wang D. G., 2020. Residential relocation and travel satisfaction change: An empirical study in Beijing, China. Transportation Research Part A: Policy Practice, 135, pp. 341–353.

Wang D., He S. J., Webster C., Zhang X. L., 2019. Unravelling residential satisfaction and relocation intention in three urban neighborhood types in Guangzhou, China. Habitat International, 85, PP. 53–62.

Chapter 5
Agent-Based Simulation for Decision-Making Support of Spatial Strategy for Large-Scale Shopping Center Development

5.1 Introduction

In this chapter we will introduce how agent-based modeling used in simulating of the interactive process of household shopping and shop location choice. The simulation model is designed to decision-making support of local spatial strategy on large-scale shopping center development. Sometimes, the large-scale shopping center development in suburban area may lead to the decline of downtown. This phenomenon real happened in some Japanese local city in last decades. Thus, what kind of planning regulations should be made to prevent the negative effects of large-scale shopping center development on downtown's vitality? And do these regulations really work? To these questions, extracting the development rules developed by the restrictions on the layout and floor space of large-scale shopping malls (hereinafter referred to as B-shops) established by the Urban Planning Law of Japan and the activation of the centers of local cities based on the law are the first steps. Based on these rules, different policy schemes are designed as different scenarios in simulation. Basing on the development of an agent-based ShopSim model, the scenarios stipulate the limitation of the location sites and floor space upper limit of different B-shops.

5.2 Agents in the Urban Space and Their Behavior Rules

5.2.1 The Framework of ShopSim Model

In order to study the influence of B-shop development rules, we used this model to design four agents. Both types of agents directly affect the physical infrastructure of the virtual city space. And it plans to develop a new B-shop by the development

Y. Ma, Z. Shen, *Strategic Spatial Planning Support System for Sustainable Development*, Advances in Geographic Information Science,
https://doi.org/10.1007/978-3-031-07543-8_5

agent, which must conform to the land zoning, development plan and planning guidelines. It is assumed in ShopSim that the Planner agent uses a Genetic algorithms (GA) approach to draw out feasible B-shop development plans and the develop agent employs Decision Table (Arentze et al. 2000) to make decision where to locate a new B-shop. In addition to designer agents and developer agents, there are two types of agents in this virtual city space. The development of the new B-shops will affect the supply of goods, market competition and the choice of household stores. Household agents will go shopping according to their personal preferences and make their own decisions. The relationship between agents is shown in Fig. 5.1.

5.2.2 *Planner Agent*

In the Shoipsim, the planner agent represents planning authorities of a local city facing the central area decline and is responsible for drawing up the regulations of B-shop developments and initiating the process to examine the efficiency of the regulations. In the Shoipsim, the planner agent represents planning authorities of a local city facing the central area decline and is responsible for drawing up the regulations of B-shop developments and initiating the process to examine the efficiency of the regulations. After the regulations for B-shops (commercial zones and floor

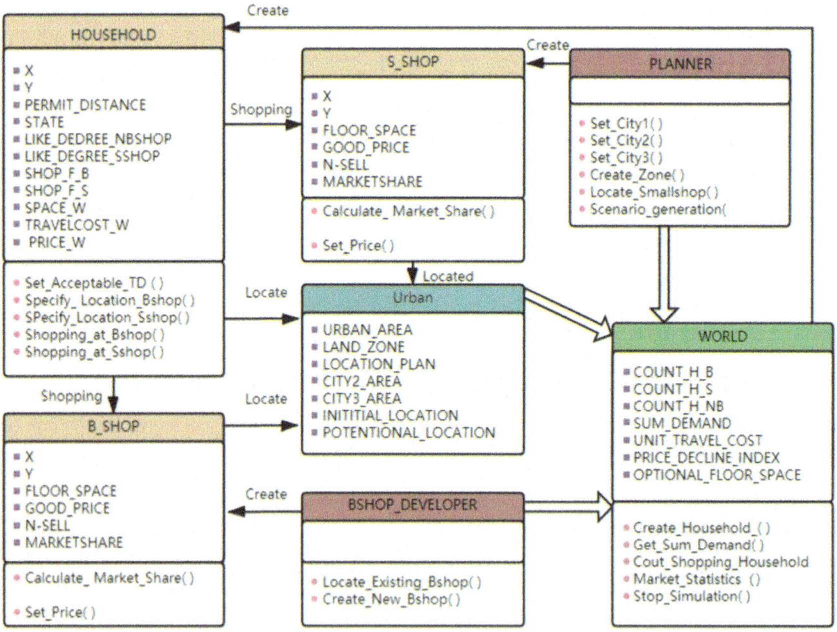

Fig. 5.1 the state diagram of agents and urban space in ShopSim

space upper-limitation of the B-shop in each zone are already set) had been formulated and the total B-shop development Quantity (in terms of floor space) for the coming year from historical statistic data is estimated, now the planner agent need to draw out the most feasible B-shop development plan which is consistent with B-shop development regulations. In this study, the most feasible B-shop development plans means that the new B-shops located according to this B-shop development plan can be accessible to as much urban area as possible. A genetic algorithm is suggested to find suitable solution from large amount of alternative B-shop development plans constrained by development objectives and regulations and designed as a decision-making tool for the planner agent.

The definitions of variables used by planner agent to make B-shop development plans are listed in the Table 5.1.

The objective of B-shop plans:

$$\max\left(\sum_{i=1}^{K} P_i - \sum_{i=1}^{k}\sum_{h=1}^{k} Z_{ih}\right) i \neq h \tag{5.1}$$

Given the objective of B-shop plans and the total development quantity (TS), try to find a set of feasible solutions to allocate TS among potential location sites, which must satisfy following planning constraints:

Table 5.1 Definitions of variables

Variable	Definition
TS	The total floor space waiting for allocation.
m	The number of existing B-shops (EB-shops)
n	The number of potential locations (sites) for B-shops
k	The number of planned new B-shops (NB-shops)
j	The jth existing B-shop
i, h	The ith or h^{th} new B-shop
d_{ij}	The Euclidean distance between the NB-shop i and existing B-shop j.
d_{ih}	The Euclidean distance between the NB-shop i and the NB-shop h.
R_i	The estimated trade area radius of the NB-shop i
S_i	The floor space allocated to the NB-shop i
P_i	The area covered by the trade circle of NB-shop i
Z	The threshold of minimum area used in GA function, denoting the trade circle of total new floor space should cover
B	The constant representing the threshold above which a shop will be called a B-shop (set according to urban planning law.)
L_i	The upper-limitation for NB-shop floor space in the potential location j
W_i	The weight of NB-shop I to get allocation of floor space (the percent of NB-shop's floor space in Total floor space).
G	The total number of initial B-shop plans (the number of chromosome)
f_g	The fitness value of g-th B-shop plan (the fitness of g-th chromosome)

$$\sum_{i=1}^{K} S_i = TS \tag{5.2}$$

The general steps of the proposed algorithm are showed in Fig. 5.2. Detail critical steps are given as following:

$$B \leq S_i \leq L_i \tag{5.3}$$

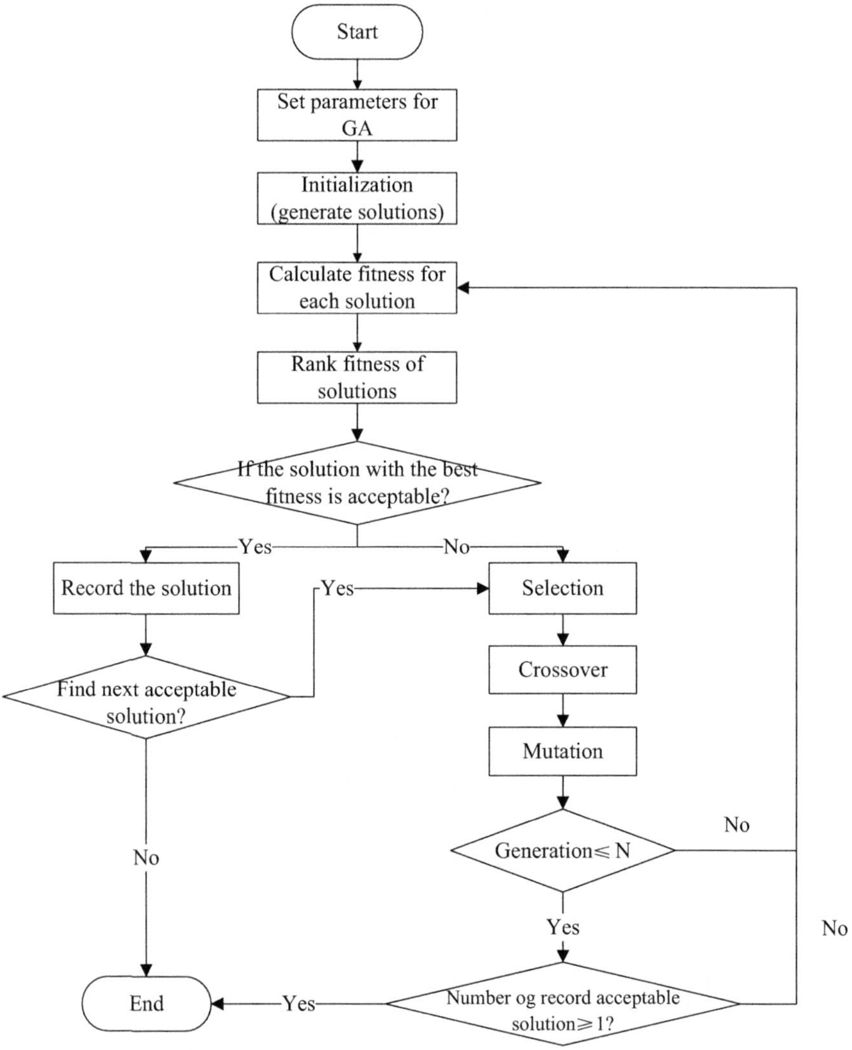

Fig. 5.2 The flow chart for planner agent to find feasible development solutions using GA

Coding New B-shop Location Solutions and Initialization

Before the search process, B-shop location solutions have to be coded into chromosomes (denoted as Chrom.), a form of binary strings. If a B-shop is located at site j then the respondent gene is unity, otherwise zero. Table 5.2 presents an example of the structure of chromosomes. Figure 5.2 illustrated the location-site allocation solution represented by Chrom.1 The search process starts with a certain number (e.g. 10) of randomly chosen chromosomes that constitute the first generation.

The floor space allocated to a new B-shop is done according to their location-sites concerning their spatial distance to each other and to all existing B-shops. The radius of a new B-shop's trade circle is first estimated using an analogous method that supposes the radius is the half of average distance to other B-shops (Fig. 5.3.). And then the percent the new B-shop's floor space occupied the total development floor space, called the weight to get floor space for the B-shop, is calculated based on its trade circle radius. Thus, the planner agent can figure out how much floor space should be allocated to the new B-shop according to Eqs. (5.4), (5.5), (5.6).

$$r_i = \frac{\sum_{h=1}^{k-1} d_{ih} + \sum_{j=1}^{m} d_{ij}}{2(k-1+m)} \tag{5.4}$$

$$W_i = \frac{r_i}{\sum_{i=1}^{k} r_i} \tag{5.5}$$

$$S_i = TS^* W_i \tag{5.6}$$

Calculating Fitness Value

This step is used to evaluate feasibility of each B-shop plan. According to B-shop planning objective, the more area new B-shop trade circles cover, and the fitter one plan is. A B-shop plan's fitness level is thought acceptable when its trade circle area is larger than Z. The fitness value of B-shop plan g is calculated out using following formula:

$$f_g = \sum_{i=1}^{k} r_i^2 - \sum_{i=1}^{k}\sum_{h=1}^{k} r_{ik}^2) i \neq h \tag{5.7}$$

Table 5.2 Generate initial solutions (Encoding)

Potential site	1	2	3	4	6	7	8
Chrom.1	1	0	0	1	0	0	0
Chrom.2	1	0	1	0	0	1	1
Chrom.3	1	1	1	0	1	0	1
Chrom.4	0	1	0	1	0	1	0
…	…	…	…	…	…	…	…
Chrom.10	1	1	0	0	1	0	0

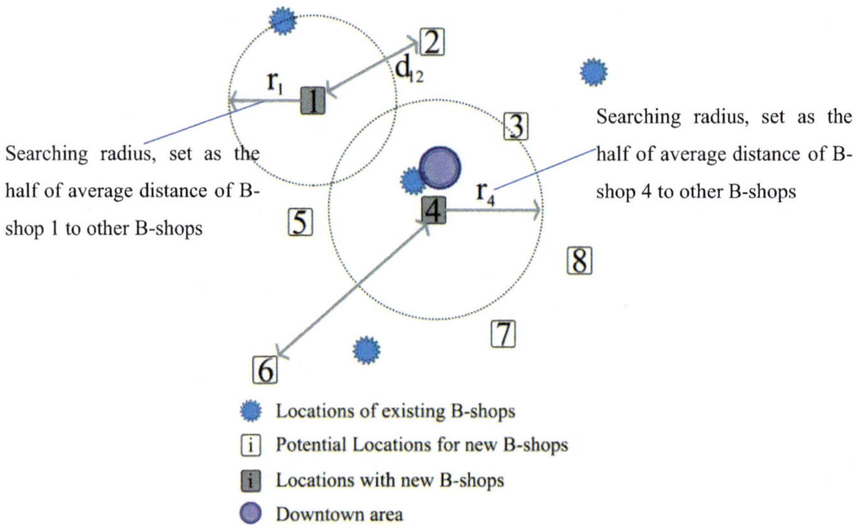

Fig. 5.3 B-shop plan indicated by Chrom.1

Evolution Operators

The method of roulette wheel is to select a parent. The probability of the selected chromosome is measured by the following equation:

$$p_g = f_g / \sum_{g=1}^{G} f_g \tag{5.8}$$

The better the chromosome, the more chances of selected. Cross is to complete the random point in the parent sequence, and exchange all the numbers in the sequence after this point after this point. For each gene of cross-modified, random number between 0 and 0.1 is generated. Since the random number is less than a given mutation rate, the gene will become one of the other two different numbers in the same accident.

5.2.3 *Large-Scale Shop Developer Agent*

Location may be very important for the development of shopping centers because it has the role of sales, market share and interests. Due to its importance and complexity, the problem itself has received a lot of attention. However, how to choose the place where the store has the best competitive advantage, is not the focus of this paper. In this model, large shop developer agents will investigate the possible position of large shopping centers based on restrictions and relative principles. When the user clicks the startup scheme button, the planner agent generates the initial setting of the scene, the developer agent finds the position consistent with the conditions described in the initial settings. In order to represent the developer agent how to find a suitable place for the new large-scale shopping centers, we adopt the Decision Table (Arentze et al. 2000) to describe this process (Table 5.3).

Table 5.3 Decision table of developer agent for large-scale shop location

Scenario	Center activation				Railway Station Development (RSD)				Neighborhood commerce			
Location is permitted in UPA	Y			N	Y			N	Y		N	
Location is permitted in land use zones	Y		N	/	Y		N	/	Y	N	/	
Floor space up-limit is acceptable	Y	N	/	/	Y	N	/	/	Y	N	/	/
Create new B-shop	Y	/	/	/	Y	/	/	/	Y	/	/	/
	R1	R2	R3	R4	R5	R6	R7	R8	R9	R10	R11	R12

5.2.4 *Shop Agent*

In addition, the store agent is divided into two types according to floor space:B-shop agent and S-shop agent. The S-shop here refers to a small-scale shop. After obtaining the development license from the designer agent, developers searched new B-shops; S-shops are concentrated in commercial areas, there is no new S-shop in this simulation.

Assume that the S-shops has the same homogeneous attributes: commodity prices and floor spaces. However, households and B-shops are different in nature. The floor space and price of the existing B-shop will be given. According to local planning regulations, the floor space of the new B-shops will be set when the developer agent opens. In fact, the new S-shop competition strategy is generally not disclosed, so the new B-shop assumes to start as a price strategy of Eq. (5.1) to compete with existing stores. Investigate the competition impact of the S-shop in price, layout, and the charm of the existing B-shop on the floor space.

Assume that small shops have the same properties, that is, the same price and size of goods. Households and B-shops however are heterogeneous. Existing B-shops' floor spaces are given exogenously. According to local planning regulations, the floor space of the new B-shops will be set when the developer agent opens. Assuming that all B-stores have the distance to decay index from the distance of the city center, and are affected by the floor space and unspected factors, given by

$$P_j = K^* EXP\left(-b^* d_{oj}\right) - Rnd(\ ) * S_j / a \tag{5.9}$$

Where

P_j is the price of the j th B-shop;
K is a constant, equal to the price of small shop in city center;
b is the price decline index, which is given exogenously;
d_{oj} is the distance of the jth B-shop to city center O;
Rnd () is a random number generated by computer, representing uncertain part of price derived from the influence of shop's floor space,
S_j is the floor space of the jth B-shop;
a is an exogenous constant, here its value is set equal to 500.

5.2.5 Household-Agents

When the household agent decides to go shopping, there will be different valuation in the distance, price or shop's floor space. By simulation, two market share of B-shop and S-shop appear as space modes in the simulated world. In order to assess the scenarios generated by the planner agent, the spatial ratio to the change of the S-shop to B-shop can be used. The decision process of household agents follows a flow chart showed as Fig. 5.4:

We use standard random tool framework (Baltas et al. 2001) to choose the household shop. In all steps, the family basically continues to compare the shopping tools for the store, choose to go shopping, unless his requirements have been satisfied. When all agents'total demands are satisfied, the simulation process will be ended. It is assumed that:

Fig. 5.4 the decision tree for household agent choosing shops

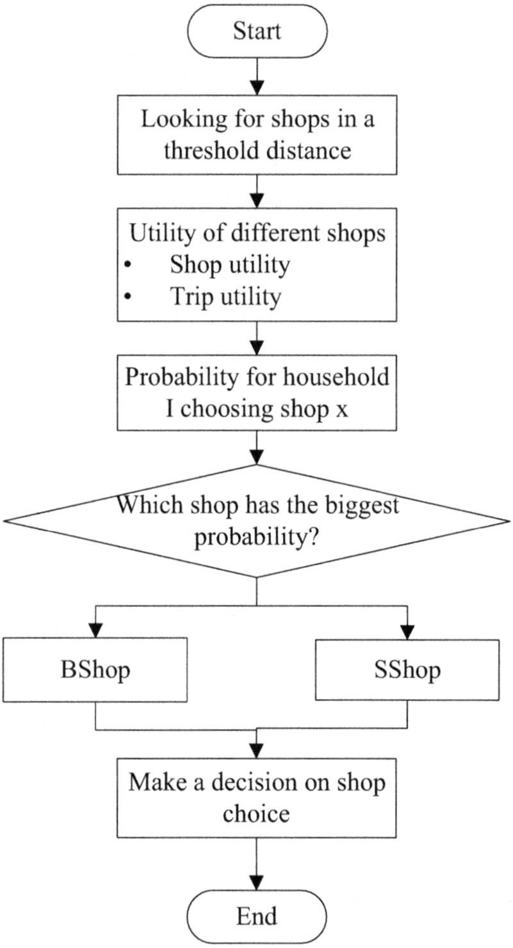

1. The goods sold in all stores are homogeneous, that is, the family will buy the same product in all the stores.
2. Each family has the same needs for goods (i.e. 50 units requirement).
3. Each step, the family wants to buy a unit of demand.
4. Some families only consider the store within a certain distance, his delta distance is γ.
5. Available stores under utility conditions. When available stores are in the same effect condition, the family will randomly choose one of them.

The utility of household i associating with alternative shop j is given by

$$U_{ij} = V_{ij} + \varepsilon_{ij} \tag{5.10}$$

$$V_{ij} = \sum_k f_j\left(X_{kij}\right) - TC_{ij} \tag{5.11}$$

$$f_j\left(X_{kij}\right) = \beta_{ik} \cdot X_{kij} \tag{5.12}$$

$$TC_{ij} = \lambda_i \cdot \alpha \cdot \sqrt{\left(x_i - x_j\right)^2 + \left(y_i - y_j\right)^2} \tag{5.13}$$

Where

V_{ij} is the deterministic part of the utility.

X_{kij} is the kth attribute describing store j presented to household i, which includes price X_{1j} and floor space X_{2j}.

$f_j\left(X_{kij}\right)$ is a function to evaluate the attractiveness of the kth attribute of shop j to household i.

TC_{ij} is a measure of the disutility of travel between site of household i and site of shop j.

β_{ik} is a specific weight of the household agent i with regarding to the attribute k of a shop.

λ_i is a parameter reflecting the attitude of the household i toward the cost of travel, here it is set equal to -1.

α is an exogenous constant that represents the unit travel cost.

ε_{ij} is the unobserved random component of utility that is used to capture uncertainty of shopping behavior.

Select the shops with the highest utility, given by:

$$\mathrm{P}\{\text{choice is } j\} = \mathrm{P}\left\{U_{ij} \geq U_{ik}, \text{for all } k \neq j\right\} \tag{5.14}$$

5.2.6 *The Interactions Between Agents*

Just as what we introduced in Chap. 4, the interactive activities between households will influence on household's decision-making. If one household chooses a shop for shopping basing on the utility comparison, this is an individual behavior. We have discussed the differences and relationships between individual behavior in a micro model and agent's behavior in an ABM in Chap. 1. Thus, Eqs. (5.10), (5.11), (5.12), (5.13), (5.14) described in Sect. 5.2.5 indicate the shopping choices of household agents at a separate level, without considering the process of interaction between agents. we take the similar logic of representing interactive process between agents in Chap. 4, in which we consider the information sharing between neighborhood and whole city. In this study, however, we define information sharing from shop agents to home agents on a global scale. Meanwhile, the information sharing among neighborhood will also be considered as a peer impact. These interactive processes will play a role in influencing agents' utility evaluation on shop choices.

Till then, it is not hard to find that the utility model for household agents choosing shops should be developed in ShopSim. In the model we use the Moore8 neighborhood to recognize one agent's neighbors. Regarding to the neighborhood relationships in an environment we have already introduced in Chap. 2 (please refer to Fig. 2.1). We define that if one agent A locates in cell i, the agents who locate in the 8 cells around cell i are A's neighbors. The agents locate in this 9-cell neighborhood area around cell i is a neighborhood. The utility of an X-type store (like an S-shop) can be enhanced by peer influence of neighbors who shop in the same type of store (any S-shop). Household agents will share shopping information such as the price, the category of goods, the environment of shopping, etc. in a neighborhood. In the meantime, this kind of information is delivered to you by a shop agent to household agents. The ubiquity of this information may appeal to household agents which were previously consumers from other stores. In this study, the concept constructs proxy variables based on the number of different types of consumers in the search area around each target store. The utility of X-type shop (e.g. S-shop) can be enhanced by disseminating information to consumer households of different types of shop now (e.g. B-shop), as a result, these households could potentially be attracted to the X-type shop.

Formula (5.15) models the development utility (the term interaction utility) resulting from the interaction between agents. The interaction utility of household i in income group g shopping at shop j is denoted as INT_{ijg}. It consists of the peer impact I_{ijg} and the information delivery D_{ijg}. In the equation, subscript x refers to the type of store j belongs to. In this study, there are just two types, S-shop and B-shop.

The peer effect I_{itg} and message delivery D_{itg} are measured in Eq. (5.16).

$$INT_{ijg} = I_{ixg} + D_{ixg}, \mathrm{x} \in \{S, B\}, \mathrm{j\,is\,type\,x\,shop} \tag{5.15}$$

$$I_{ijg} = k_g \left(N_{ixg} - N_{iog} \right) / N_{ig}$$
$$Di_{jg} = d_g \left(N_{iogd} - N_{ixgd} \right) / N_{igd} \quad (5.16)$$

where

N_{ixg} is the number of i's neighbors who are in g^{th} income group and shop at x type of shop.
N_{iog} is the number of those who shop at the other type of shop.
$N_{ig}$is the total number of i' s neighbors in g^{th} income group.
k_g is a scaling factor reflecting household agents' subjective reaction to such influences, which is a constant.

The symbol with subscript d means that the respective quantity is calculated in the search area with radius d around store j. The interaction utility generated by the number of shops of any type in a neighborhood works in opposite ways through I_{ijg} and D_{ijg}, respectively, so that the overall interaction utility takes on a waveform.

After considering the interaction between agents, we change the utility function defined in Eqs. (5.10), (5.11), (5.12), (5.13), (5.14), (5.15), (5.16), (5.17). At the beginning of the simulation, the initial utility value is calculated from Eq. (5.10). Then interactions will take place and Eq. (5.17) is used in following simulation.

$$U_{ijg} = \sum_{n=1}^{m} \mu \beta_{ign} X_{ijn} + \mu INT_{ijg} + \varepsilon_{ijg}$$
$$\varepsilon_{ijg} = a - b\left(\ln(-\ln\theta)\right), \theta \in \left[-\theta_g, +\theta_g\right] \quad (5.17)$$

Where:

U_{ijg} is the utility of household i in income group g received from shopping at shop j.
X_{ij} is a vector of observable explanatory variables of shop j produces to household i. These variables include the travel cost of household i to shop j (the detailed information of travel cost will be introduced in Sect. 5.3), price of goods and floor spaces of shop j. The subscript n refers to the dimension of the vector (number of variables).
β_{ig} is a vector of respective coefficients to the variables of shop j.
μ is a constant, which is generated following normal distribution with mean of 1 and standard deviation of 0.5.
ε_{ijg} is a unobserved random that contribute to the utility. It follows Gumble distribution and can be generated using a random number following uniform distribution.
θg is a pre-defined range that represents the maximal magnitude of possible internal differences within the income group g^{th}.
a, b are constants that set as 0.5 and 2 respectively.

5.3 Modelling the Transportation Mode in the Shop-Selection Model

The same with household agents choosing shops, household agents also choose different transportation modes to shopping. Furthermore, the transportation modes of household agents have a great influence on the shop-choice of agents. For instance, one household agent wants to go shopping in UPA while he/she lives in CA, but there is no bus directly arrive at the shop that locates in UPA. For this reason, he/she needs to drive car or take taxi. If these two transportation modes are not available to him/her, he/she needs to change a shop for shopping. This example shows an ordinary phenomenon that when we make a shop choice for shopping, we usually will consider the transportation mode too. In a word, the choice of shop is indeed an interactive result of shop choice and travel mode choice.

Existing researches proved that many factors influence travelers' choice of transportation mode and route (Ortúzar and Willumsen 2001). An empirical study shows that just is the combination of time and distance can explain about 60–80% percent of the variations in route changes (Outram and Thompson 1978). For this study, we suppose that household agents will evaluate on the costs of transportation modes before making-decision on a mode choice. To represent the costs of transportation modes, a generalized cost is defined as Eq. (5.18):

$$X_{ijm} = TMCost_m = DC_m + T_m + \frac{hD}{S_m} \tag{5.18}$$

Where:

$TMCost_m$ is the generalized cost for transportation mode and it incorporates the effects of distance, time cost and monetary cost of a shopping trip.

D is the travel distance.

C_m is the unit monetary travel cost for mode m.

S_m is the average speed associated with the travel mode.

T_m is parking fee.

h is a weight used as a scaling factor, which follows normal distribution with mean of 1 and standard deviation of 0.5.

Basing on Eq. (5.18), the utility of choosing different transportation modes can be evaluated by Eq. (5.19). The same with shop choices and the residential location choices, household agents may evaluate on the utility between alternative transportation modes before making a final choice. As shown in Eq. (5.19), the utility U_m is formed by V_m and a constant ε_{ijm}. The V_m is further determined by a random coefficient $\mu\beta_{igm}$ and X_{ijm}. The X_{ijm} stands for the generalized cost of household agent i to shop j by transportation mode m. Coefficient $\mu\beta_{igm}$ is a random coefficient, which is used to explain the cost among household agents. μ is generated according to normal distribution, with an average of 1 and standard deviation of 0.5. The value of β_{igm} is determined based on g^{th} income group and different travel mode m. There is also a random element γ_{igm} stands for the utility variations of household agent i in g[th] income group choosing transportation mode m due to other unobserved factors. As an independent, codistributed random element, it follows Gumble distribution generated.

$$U_m = V_m + \varepsilon_{ijm} = \mu\beta_{igm}X_{ijm} + \gamma_{igm} \tag{5.19}$$

$$P_m = \exp(V_m) / \sum_{m=1}^{M} \exp(V_m) \tag{5.20}$$

Household agents choose travel mode according to the shopping utility of the replacement shop and the generalized travel cost under the replacement mode. The processes are illustrated in Fig. 5.5. In the figure, the logit (ML) model based on utility values simulates the probability of travel mode selection (P1, P2, P3 and P4) and the probability of the shopping (Pb and Ps).

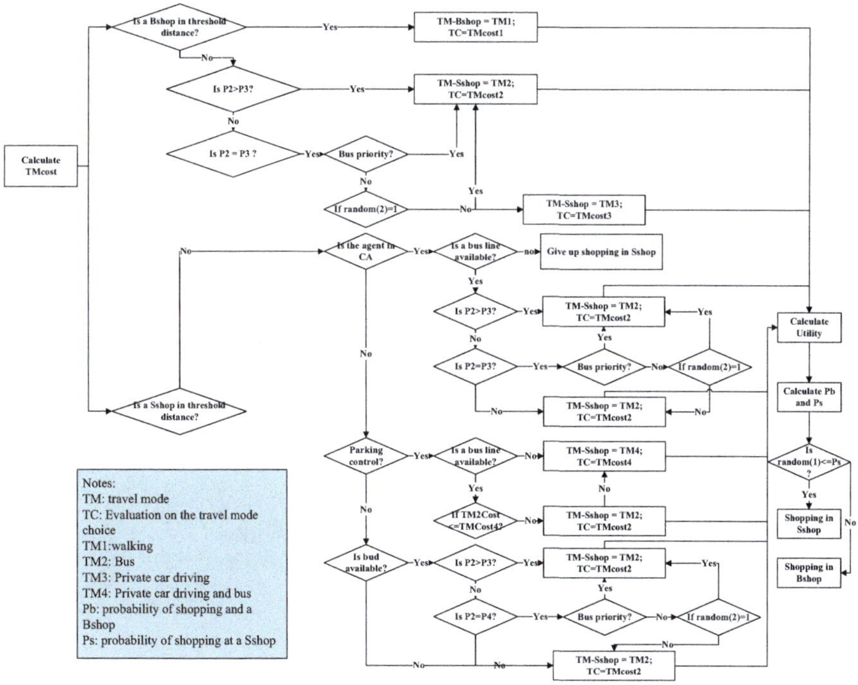

Fig. 5.5 The interactive process of household agent make decision on transportation mode and shop choice

5.4 Sensitivity Test and Validation

5.4.1 Initial Settings

The ShopSim model is developed on Netlogo platform, the Fig. 5.6 shows the interface of ShopSim model. as shown as Fig. 5.6, the blue dot in the CA area is agents of S-shop and the red dot in the UCA area is an agent of B-shop. The environment of the model is showed by Fig. 5.7, which is designed on the City Planning Law of Japan. The detailed spatial features of this environment have been already illustrated in Chap. 2. As shown as Fig. 5.7, the urban area is divided into CA, UPA and UCA respectively. Household agents are distributed on this environment and different color in Fig. 5.7 represents different household density, the darker the cell color is the higher the household density is in the cell.

There are totally 1600 household agents living in this virtual city. Households are grouped into three income levels, which are Rich, Middle Class, and Poor. We assume that percentages of population in the three income levels are 20%, 60%, and 20% respectively. We also assume that all households have cars.

In all, there are 1600 household agents in the virtual city. Families are divided into three income groups: Rich, Middle Class, and Poor. We assume 20%, 60% and 20% of the population at the three income levels and all families have cars.

As shown in Fig. 5.8, there are 17 S-shops in CA district and 1 B-shop in UCA district. The floor space of B-shop was initially set at 20,000 m^2, which was within the floor spaces of the B-shop as stipulated in planning regulations. Due to the

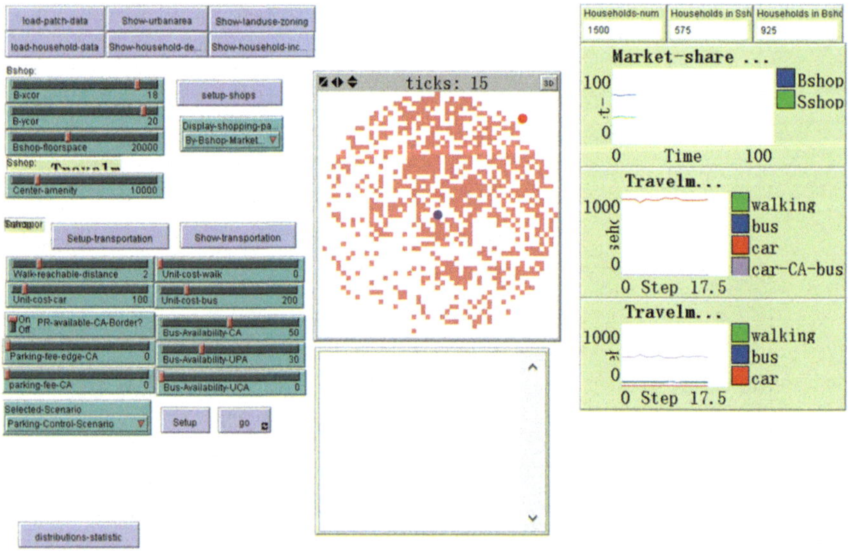

Fig. 5.6 the interface of ShopSim

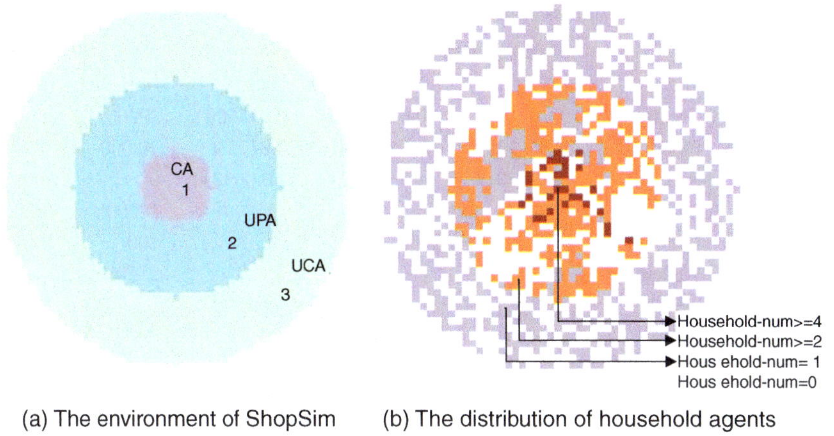

Fig. 5.7 The environment and household agents in ShopSim model. (a) The environment of ShopSim. (b) The distribution of household agents in ShopSim

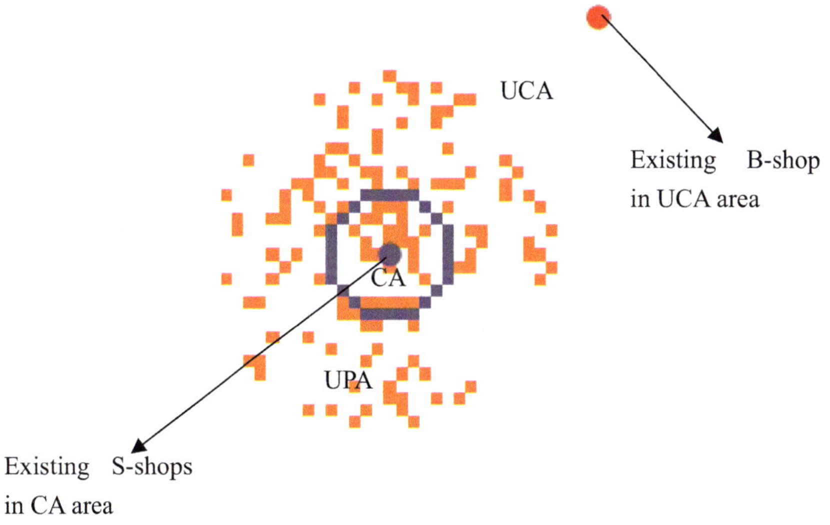

Fig. 5.8 Shops in the virtual urban

regulation of National Survey of Price (www.stat.go.jp), the small-scale shop's floor space is less than 450 m^2. Thus, we set the floor space of S-shops as 300 m^2.

Before took the model for sensitivity test and validation, the initial conditions of ShopSim should be set first. The Table 5.4 shows the initial condition of the model. Within them, the parameters for shops and travel modes are set basing on the researches of Muromachi et al. (1990) and Hanaoka et al. (2000). The price is about Japanese Yen. The values' number note as "Dynamic" means that the value will change during simulation. The parameters of household interaction are calculated basing on the average values of Mid class households.

Table 5.4 The initial conditions of the ShopSim model

Shop Utility		Average of group β			X_{ij}	
		Low	Mid	High	In B-shop	In S-shop
Shops	Goods price	0.146000	0.0730000	0.014600	JP200	JP300
	Floor space	0.002466	0.0122300	0.012230	20,000	300
Travel modes	Parking fee	−0.038443	−0.0192217	−0.003844	0	600
	Bus cost				JP200	JP200
	Car cost				100	100
	Walkable distance	–	–	–	2*500 m	
	Bus availability	–	–	–	50%(CA), 30%(UPA) and 0%(UCA)	
Household interaction	Impact of neighborhood	0.023675	0.0236746	0.023675	Dynamic	Dynamic
	Impact of information delivery	0.023675	0.0236746	0.023675	Dynamic	Dynamic
Attitude to policy		1.000000	1.0000000	1.000000	–	Dynamic
ε_{ij}		−11.102000	−11.1020000	−11.102000	10,000	10,000

5.4.2 *Sensitivity Test*

In Sect. 5.4, the model has been illustrated very detail, but before we took the model for the practice of decision-making support, model test and validation are necessary. First, a sensitivity test will be took to prove all the policy parameters we designed in the model is sensitive in reflecting shopping behaviors of household agents. Figure 5.8a–d shows the results of sensitivity test. For the test, each parameter we designed in the model has been tested respectively. This paper demonstrates whether a parameter can effectively affect the shopping behavior of family agents, all parameters are valid in affecting the agents' shopping behaviors when they work together. In the following we will introduce the test results of parameters one by one.

1. Unit transportation cost(α)

Figure 5.9a shows the relationship between store sales and unit travel cost. It is assumed that all stores are accessible to families (accessible distance = 30). The lower the unit travel cost, the more likely the family is to go shopping in large stores outside the center, and the less goods the center sells. Obviously, if the unit travel cost is equal to 0, the market share of the central store will be reduced to 0.

2. Reachable distance (γ)

Figure 5.9b shows the relationship between reachable distance and store sales. Reachable distance is a variable that represents how far a household can go shopping. In the simulation, when the entry distance is less than 15 (500 meters is 1),

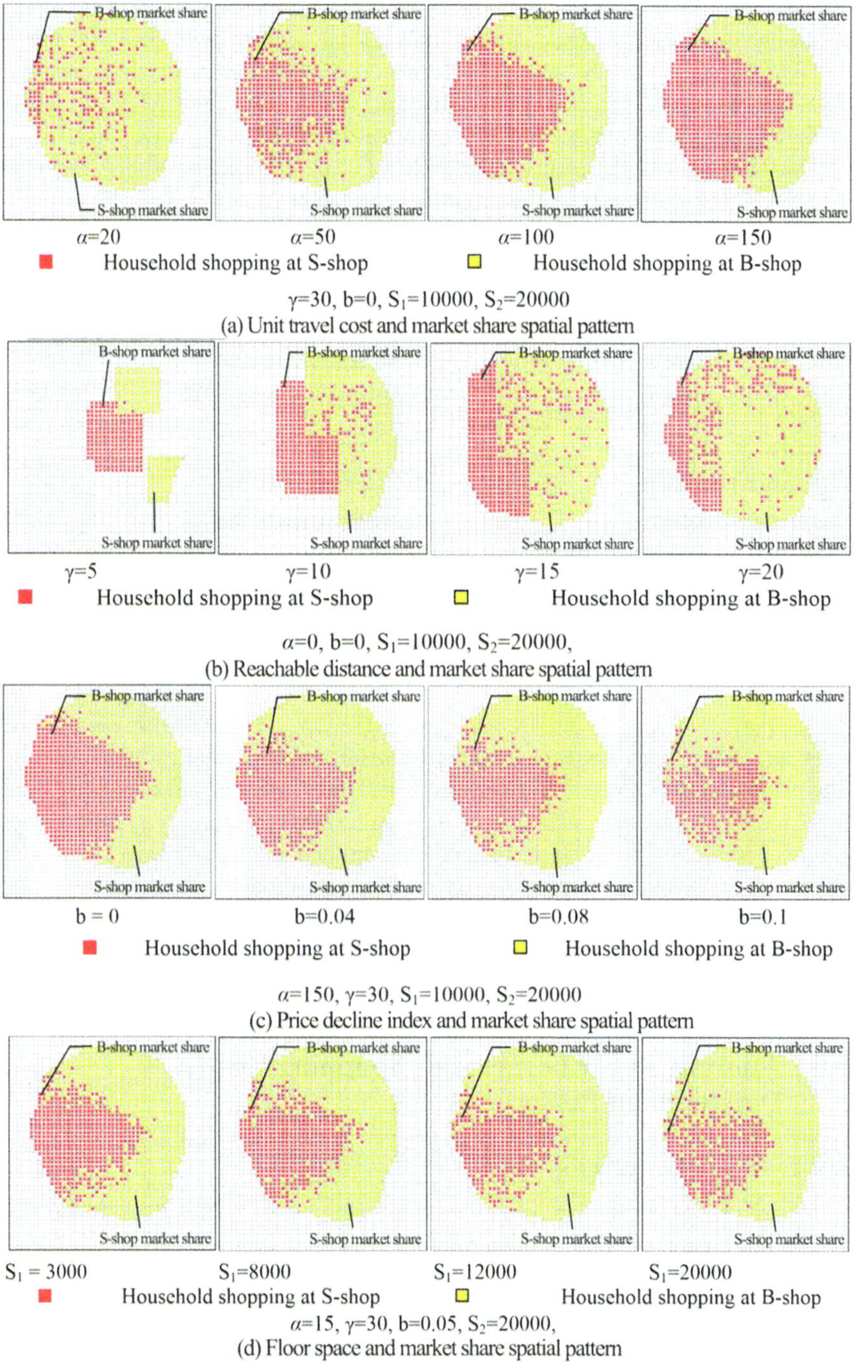

Fig. 5.9 Sensitivity tests of parameters. (**a**) Unit travel cost and market share spatial pattern. (**b**) Reachable distance and market share spatial pattern. (**c**) Price decline index and market share spatial pattern. (**d**) Floor space and market share spatial pattern

the sales of both B-shops and S-shops increase. After this, S-shops' sale amount dropped sharply. This phenomenon shows that households will choose to shop at large-scale shop if they are accessible and do not have to pay for travel cost. This is easy to be understood that usually a big shop's goods are richer in varieties than small shops, and meanwhile the price would be cheaper that small shops too. Thus, household agents make decision on shop choices actually is an equilibrium process between different utility parameters.

3. Price decline index (b)

Figure 5.9c shows the relationship between price decline index and store sales. In the simulation, under the condition of keeping other parameters unchanged, the price decline index shows an approximate linear relationship with the sales of stores. The larger the index, the lower the commodity price of store B and the lower the sales of store S. This is consistent with our experience. The low price of b-shop can usually attract more families.

4. Floor space (S_j)

Figure 5.9d shows that the sales performance of central stores is also affected by the floorspace of B-shops. The larger floor space means that B-shop can provide goods and services to families at a relatively low price, which also indicates that floor space may affect prices in unobserved ways. For example, a households' shopping experience can allow them to associate a larger floor area with a better shopping environment, a rich selection of goods and lower prices. Therefore, increasing the large floor space may cause S-shops to lose customers.

All the parameter tests above are made under the same spatial features of the environment. During each single test, other parameters are kept stable at initial settings. The parameter tests made in this section proved the validation and sensitivity of the four kinds of parameters in simulation. Research shows that travel cost, store accessibility, commodity price, store area and other influencing factors have a significant impact on households' shopping behavior. Thus, the model is sensitive in simulating the performance of city center commerce affected by the B-shops in outside CA area.

5.5 Analysis on the Large-Scale Shopping Center Development Policy

5.5.1 Case Study Area

To test the model, we took a Japanese local city as a case study area. Firstly, the planning policies regarding to large scale shopping center development have to be collected. In practice, each city has different regulations on the development of B-shops, such as possible site selection and floor area restrictions, which are explained in different forms. In this study, the regulations on the location of B-shops refer to the information described in Urban Planning Law in Japan. We also took

Table 5.5 site selection of B-shops in Kanazawa City by law

Location candidate sites	Requirements on candidate sites	Upper limit of floor space (m^2)
Central area	CBD	No limit
	Improvement areas along main road	20,000
	Other improvement areas	3000
Railway Station area	Areas along the main road connecting to station and other major transport facilities	10,000
Cultural preservation zone	Areas along main road	3000
	Other areas	1000
Sub central area	Areas along main road	5000
	Other areas	1000
Neighborhood commercial areas	Areas along main road	3000
	Other areas	1000
Residential areas	Areas along main road	3000
	Other areas	1000
Industrial areas	Areas along main road	3000
	Other areas	1000

Kanazawa City as case study are to collect local regulations on the development of B-shop, which were issued in 2002 (see Tables 5.5, 5.6). In the urban planning area of the city, according to the planning zoning types defined in the Commercial Environment Planning, the location restrictions and the upper limit of floor space of B-shop are also different. Table 5.5 shows the floor space upper limitation of B-shops on different land use zone and location candidate sites basing on law. In the following simulation, regulations will be designed as possible policy scenarios and launched in the virtual city. Table 5.6 shows the spatial control of B-shop development in different land use zones.

5.5.2 *Simulating the Spatial Patterns and Market Shares*

Basing on the policies we introduced Tables 5.6, 5.7, we use ShopSim model to simulate the spatial patterns and market shares during the process of B-shop development. In the simulation, the market share of the store was defined as the unit score in which the rate of household agents shopping in the store exceeded 80%. Although different household agents regarding to a same shop may have different estimation of utility, if there is not interactions between households the market share is easy to be figure out at the first step of simulation. That is why we need a random component in the utility model, which helps us the observe the different spatial patterns of market sharing. This is a very important point of using agent-based model to analysis individual behavior related macro spatial pattern. The random component improves the ability of utility model in reflecting the uncertainty of individual

Table 5.6 Planning regulations on location of the B-shop

Urban planning area	Land use zone	Permitting state
Urbanization promoting area	1st low-rise exclusive residential district	X
	2nd low-rise exclusive residential district	
	1st mid-high exclusive residential district	
	2nd mid-high exclusive residential district	
	1st residential district	
	Exclusive industrial district	
	Commercial district	O
	Quasi-industrial district	
	Industrial district	
	2nd residential district	
	Quasi-residential district	
	Neighborhood commercial district	
Urbanization control area	–	▲
White land	–	O
X	B-shops are not permitted to locate in these land zoning district	
O	B-shops can be permitted to located in these land zoning district	
▲	In principle any development is prohibited in urbanization control area.	

Table 5.7 Comparison of sale amount in real and virtual center

Mesh size	Shop (number)	Floor space (m^2)	Sale amount	Market share (%)
Reality (1 km^2)	S-shop (2006)	192,445	186(billion yen)	89
	B-shop (4)	26,483	22.4(billion yen)	11
Virtual (0.25 km^2)	S-shop (36)	10,800	55,475(demand)	90
	B-shop (1)	1500	6025(demand)	10

decision making, and hereby, reflecting the complexity of individual behavior. The simulated results of spatial patterns of market sharing between B-shop and S-shop is showed in Fig. 5.10. However, when S-shop and B-shop are in the same place, the market share boundary will not appear, and a random spatial pattern will appear (the second picture on the left of Fig. 5.10). In the case of B-shop and S-shop are in different urban planning zones, as shown as the third and fourth pictures show from the left of Fig. 5.10, the smaller the random component is, the clearer the market sharing boundary is. Thus, the differences in random component between household agents will cause different spatial patterns of market sharing and the smaller random component is enough to reflect the impacts of household agents' shop choices. Therefore, the S-shop and B-shop of the utility model adopt the random value 500 for further simulation.

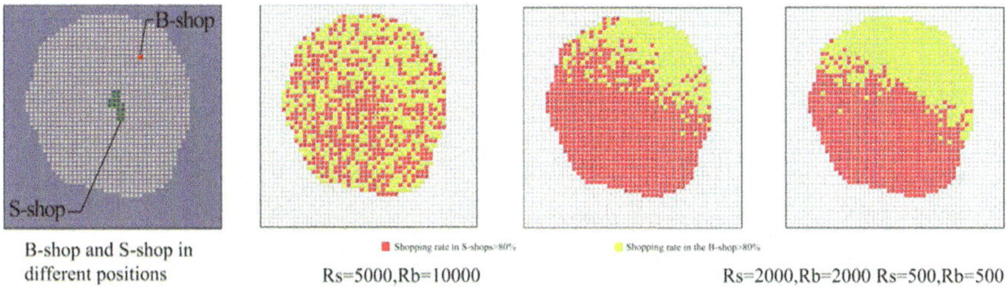

Fig. 5.10 the random setting and related spatial patterns of market share

5.5.3 *The Hypothetical Urban Space*

In order to simulate B-shop development regulations effect on inner city regeneration by ShopSim model, we designed three policy scenarios. According to relevant regulations, combined with the site selection and building area ceiling of B-shops, the corresponding policy scenarios is designed. These scenarios should be executed in the virtual city, which is the main part of ShopSim urban space.

This s hypothetical urban space is the environment for accommodating household agents' shopping behaviors. The detailed information of how to create this environment has been introduced in Chap. 2. As described in Chap. 2, this environment consists of 2500 batteries (50 × 50) with a size of 500 m × 500 m for each cell. The urban space consists of a central city and two neighboring cities. The model assumes that the central virtual city follows the typical characteristics of Japan City Planning Law and shows typical spatial characteristics of Kanazawa City. In this virtual city, there is a traditional business center located in the city center. The urban planning is divided into Urbanization Promotion Area (UPA) and Urbanization Control Area (UCA). The total area of the urban planning area of this virtual city is 1230 cells. Land use zones are defined within UPA and controlled in UCA. In the ShopSim model, all the planning information is controlled and known by planner agent, the planner agent will set the information to the cells in UPA.

The virtual city and detailed land use zoning in CCA is showed by Figure 5.11a, b. Types first-sixth in Figure 5.11b are land use zones where B-shops are allowed to be located.

5.5.4 *Policy Scenarios*

For intuitively visualize the impact of the new B-shop development on the vitality of city center commercial, the shopping behaviors of household agents will be simulated for producing the spatial patterns of market shares between different shop agents in the urban space. The simulation process follows a decision-making process that we introduced in Chap. 1. The details are showed as following:

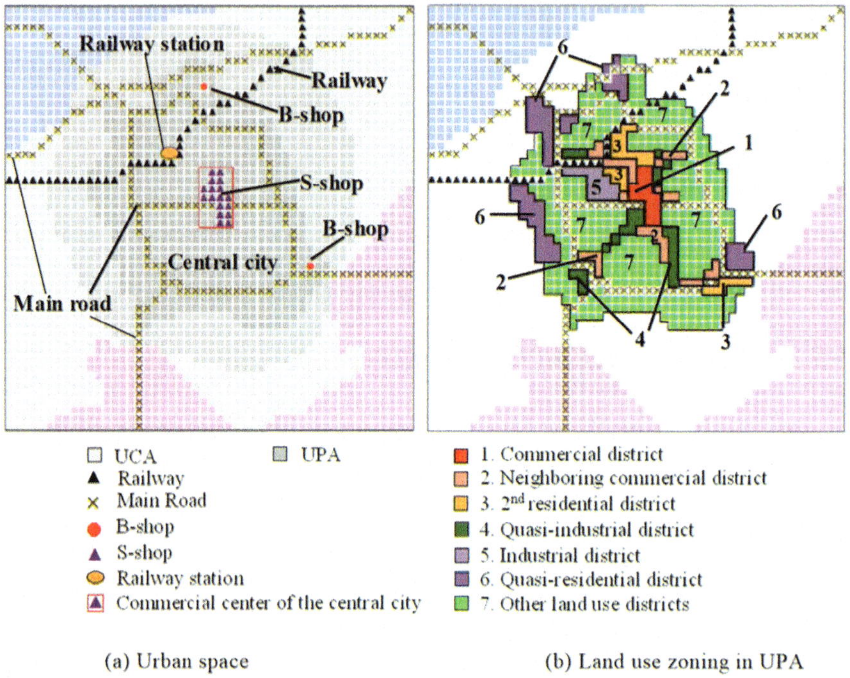

Fig. 5.11 The virtual city and land use zones in UPA. (**a**) Urban space. (**b**) Land use zoning in UPA

1. Policy scenario should be firstly defined by the user of ShopSim model.
2. Spatial features and initialized spatial parameters will be set by planner agent regarding to the defined scenario.
3. Create agents in the urban space, including the existing S-shop agents and S-shop agents. The household agents that created in the urban planning area are distributed to the CA and UPA.
4. The shop developer agent develops the new B-shop outside the commercial area of the city center according to the defined scenarios.
5. The original value of parameter set by user including unit travel cost, the threshold for household agent choosing shops.
6. Households searching for shops within the threshold distance and choosing transportation mode basing on utility evaluation, and then make decision on shop choices according to the final utility the shop makes to him/her.

In accordance with the development provisions described in Sect. 5.5.2, three scenarios were developed to simulate the positioning of the new B store in a virtual city:

1. Center Activation (CA): to reverse the downward trend and revitalize the central commercial, this policy scenario targets to encourage large-scale shop to choose their location in the city center commercial area without a floor area cap, but to severely limit the development of large-scale shopping center outside the original center commercial area.

2. Railway Station Development (RSD): it is proved that a combined spatial function in railway station is good for improving the activity of the land use plots around the station. In this scenario we hereby target to develop the railway station area as an integrated business district and locate the B-shop in the neighborhood land use plot of the station with an upper limitation of 10,000 m^2.
3. Neighboring Commerce Promotion (NCP): in order to improve the convenience of the community, B-shop is encouraged to be located in neighboring commercial area, and the maximum limit is 3000 m^2.

5.5.5 *Model Test*

Such as the utility model for household agents choosing transportation mode and shopping shops, the key parameters that would affect household agents' utility evaluation are unit travel cost (c), reachable distance(γ) and floor space (S). In order to verify the operation characteristics of Shopsim model, parameter sensitivity analysis has already been introduced in Sect. 5.4.2. The simulation is executed under the following parameters. The households in a unit should be homogeneous and are treated as one agent. Since there are 1230 units in the virtual city space, so in the simulation there are 1230 household agents. In one month, each household agent are defined to shopping 50 times. There are 2 B-shop and 17 S-shop existing in the urban area. According to the floorspace stipulated in planning regulations, the floorspace of the first B-shop is 10,000 m^2 and the floorspace of the second B-shop is 20,000 m^2. From the National Survey of Price (www.stat.go.jp), small shops have a floorspace of less than 450 m^2. The floorspace of small shops in Kanazawa city center should be lower than the floorspace specified in the National Survey of Price, and should be less than 300 m^2.

For ShopSim model calibration, the one-way cost is 20 Japanese Yen, the average ticket price of 500 m of carriage. Therefore, the entry distance is set to 30 cells, representing 15,000 m. Commercial Statistics Survey in 1985, Digital National Information (http://nlftp.mlit.go.jp) is used for this testing. Figure 5.12a shows the spatial distribution of shops in Kanazawa city in 1985. The meshes with more than 150 stores is identified as the city center and can accommodate 2006 S-shops and 4 B-shops, with a total area of 192,445 m^2 for S-shops and 26,483 m^2 for B-shops. The central area is mapped into 36 units to accommodate 36 S-shops (300 m^2 per S-shop) and 1 B-shop (1500 m^2). The total area of S-shops' floor space is 10,800 m^2 and it is 7.2 times bigger than B-shops' total area of floor space. This character is almost the same with the situation in real Kanazawa City. Meanwhile, the household agents are set up the same as the parameter examination (the Sect. 5.4.2). Figure 5.12 shows the simulated shop locations in virtual urban space and the real distribution of shops in Kanazawa City. It is showed that although there is no obvious shop sharing boundary between S-shops and B-shops, the performance of S-shops on market sharing is much better than B-shops. This phenomenon probably due to the advantages of S-shops in number and floor space.

The simulation results are further listed in Table 5.7 to show the statistic relationships between the actual data and simulation results, which are consistent. We can conclude that ShopSim can be used as a decision support tool to simulate the market shared space model between B-shops and S-shops.

5.5.6 Comparison of Different Scenarios

Three scenarios were simulated using the same parameters used in parameter sensitivity tests. The location of existing shops is shown in Fig. 5.12, which is similar with the clustering situation of B-shops in the east and south of Kanazawa City. Figure 5.13 show the spatial effect of the CA scenario. The spatial model in Fig. 5.13 is expressed by market share as purchase rate. From the figure we can see that there is a market sharing boundary between center shops (including S-shops and the new B-shop) and existing B-shops. However, since S-shops and the new B-shop are located in central city, their market sharing appear a random spatial pattern. In the simulation process, after a new B-shop is developed in the city center, S-shops will face fierce competition, which will lead to the loss of a large part of s store's market share.

From Fig. 5.14, it shown that the larger the new B-shop is, the higher the market share of the core store is. However, during this process the S-shops' market share decreased. This phenomenon shows that the CA scenario has effectively improved the market performance of global central shops, but the effects is negative to the center S-shops. In this case if the negative effects are resulted from the B-shop's scale, the upper limit of its floor space is necessary, otherwise the B-shops will attract more and more customers give up shopping in S-shops.

In order to make the simulation results could be compared between different scenarios, the new B-shop developed by the developer agent in each scenario follows the same floor space limit about 3000 m^2. In scenario RSD and NCP, the competitiveness of S-shops is lower than that in scenario CA, but market statistics show that RSD and NCP have very little impact on center commerce revitalization

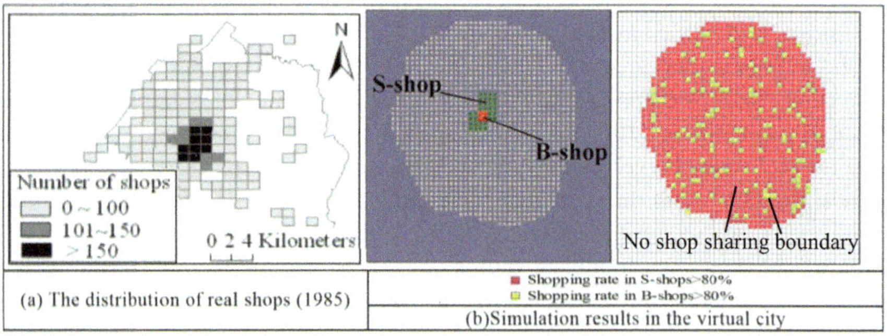

Fig. 5.12 the comparison of real shops' distribution and simulation results. (**a**) The distribution of real shops (1985). (**b**) Simulation results in the virtual city

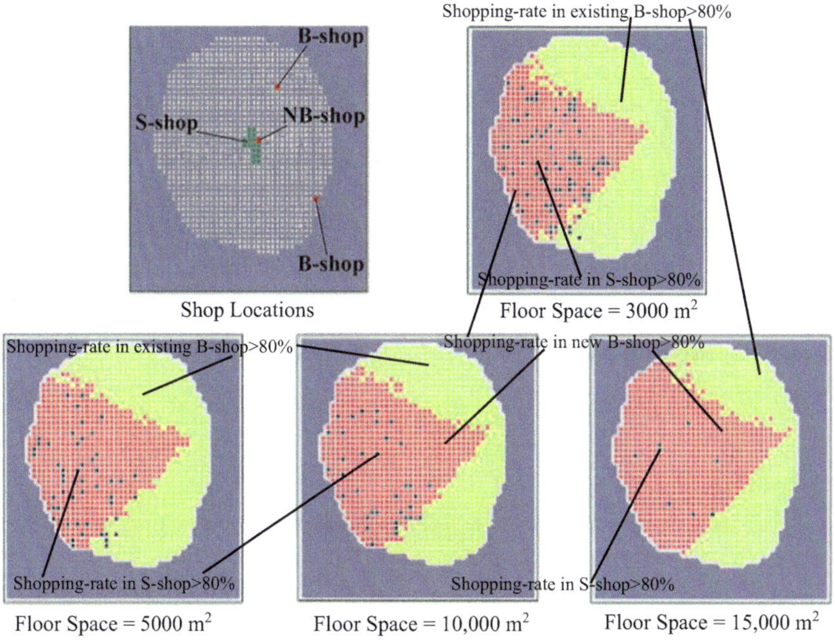

Fig. 5.13 The spatial pattern of purchase rate in CA scenario

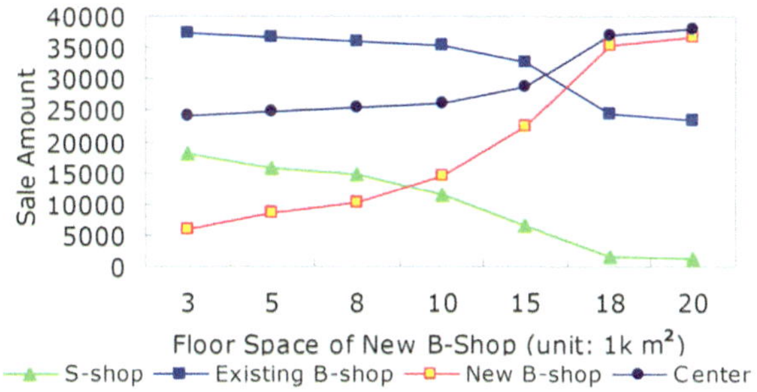

Fig. 5.14 The sale statistics in CA scenario

(Fig. 5.15). Meanwhile, the market share loss of S-shops caused by new B-shop in scenarios RSD and NCP is more than that in scenario CA, so CA scenario may be an effective measure to improve the commercial activities of the center (as shown in Fig. 5.16). Of course, this conclusion is basing on the condition that the floor space of the new developed B-shop if controlled under a reasonable limit. In the Kanazawa urban master planning, both CA and NCP schemes are accepted in the planning concepts, which may help resolve the conflict between the central commercial center revitalization and city expansion. If the capabilities of the model can be extended to accommodate complex realities, ShopSim can be a promising tool for analyzing the dynamic processes associated with B-shop planning.

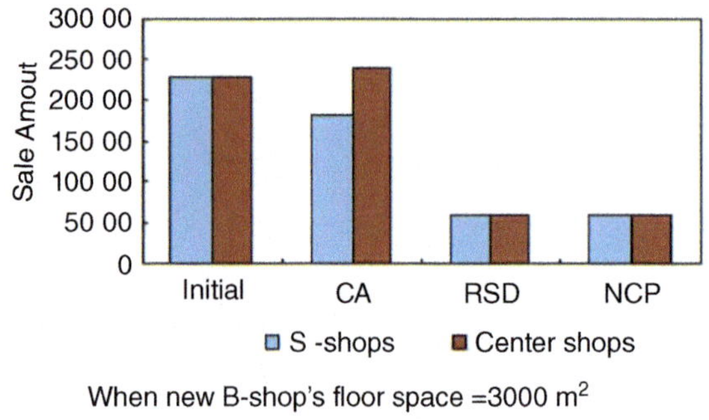

Fig. 5.15 Central commercial area shops' sales amount statistics

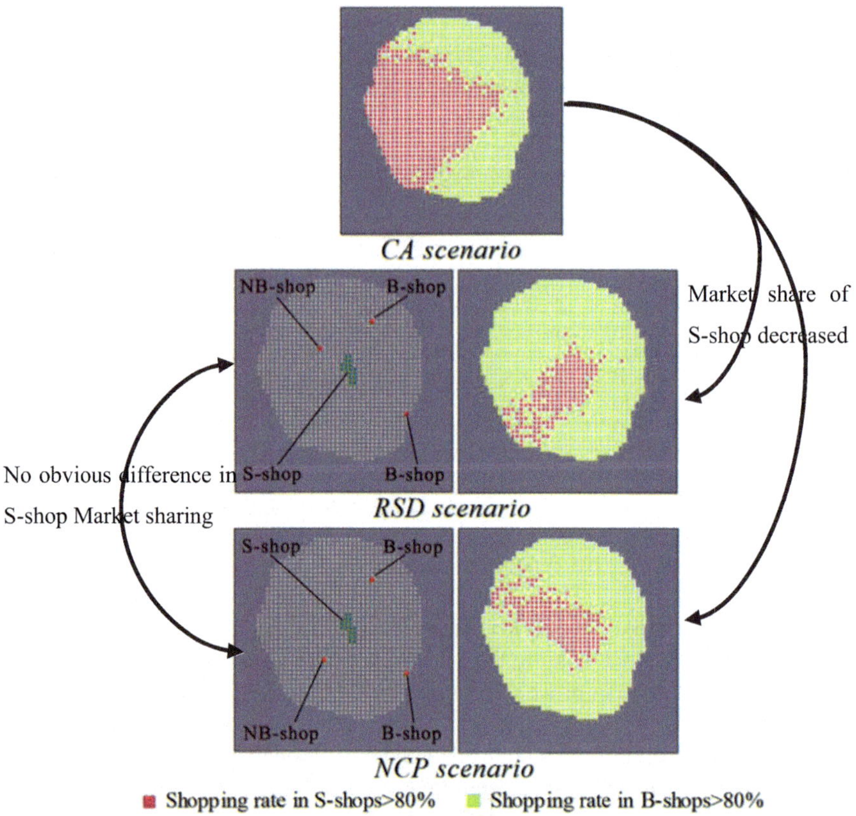

Fig. 5.16 Spatial distribution of purchase rate in different scenarios

5.6 Discussions

In this chapter an agent-based ShopSim model is developed basing on utility theory. In ShopSim, a virtual urban space is created as the environment of the agent-based model. Due to the research purpose of this work that the model will be utilized to support decision-making on the large-scale shopping center development strategy, so that although the environment is virtual, it should be able to reflect the spatial features of research area. According to real local cities are facing decline in Japan, we took Kanazawa city as a case study area to create the virtual space. The same with what we introduced in previous Chapters, when an agent-based model is going to be employed for decision-making support, the environment, agents and interactions between agents and environment should be a complex system as reality. This is the reason why we carefully prepare the virtual space to represent local urban planning policies and spatial features.

Furthermore, regarding to the agent behaviors such as the decision-making process of planner agent, the development process of B-shop developer agents, household agents' choices for shop and travel mode, all are designed as the behaviors that regulated and affected by urban planning regulations. This is very important for policy analysis or decision-making support. To fulfill this purpose, many parameters are designed in the model and some of them will be planning parameters, such as the limit of floor area of different shops, the cost of different travel modes. These parameters are combined to the traditional utility model, in this case, parameter sensitivity analysis is needed. We need to firstly prove the sensitivity and validation of the parameters on simulating agents' behavior choice. Additionally, the interactions between household agents regarding to B-shop development policies should be considered to. We use the proportion of household agents in a neighborhood choosing one kind of shop type to reflect the information delivery between household agents. This design extended the utility model to a utility model considered interaction. Although the information delivery between individual is complicated and hard to be simply represent be a ratio, it is necessary and important for agent-based policy analysis. In the future work this process needs to be further studied and tested.

Different from the simulations that target to answer the effectiveness of a planning policy, this work targets to answer what kind of policies would be helpful for revitalization of central city commerce. In this case, policy scenarios are needed. The scenario analysis shows different spatial pattern of market sharing during the development of new B-shop. According to these three scenarios, the influence of new B-shop on downtown commercial environment is visualized, which shows that developing new B-shop in downtown is an effective measure to improve downtown commercial activities, provided that the area of B-shop's floor space is reasonably limited. The simulation results and relevant policy parameters will be a good reference to local policy making.

References

Baltas, G., Doyle, P., 2001. Random utility models in marketing research: a survey[J]. Journal of Business Research, 51(2), 115–125.

Hanaoka, K., Kondo, A., Hirose, Y., and Hagioka, J., 2000, "Gaps between Consumers and Store Managers in Evaluation of Shopping Complex", Journal of the City Planning Institute of Japan, 35 157–162 (in Japanese).

Muromachi, Y., Harata, N. and Ohta, K., 1990, "An Impact Analysis of a Large-Scale Shopping Center Development in the Metoropolitan Suburbs", Journal of the City Planning Institute of Japan, 25 487–492 (in Japanese).

Ortúzar, J. D. Willumsen, L.G, 2001, Modelling Transport. John & Wiley; 3rd edition.

Outram, V.E. and Thompson, E., 1978, Driver route choice-behavioural and motivational studies. Proceedings 5th PTRC Summer Annual Meeting, University of Warwick, July 1977, England.

T. Arentze, A. Borgers, H. Timmermans, 2000, A knowledge-based system for developing retail location strategies, Computers, Environment and Urban Systems 24(2000), pp. 489–508.

Chapter 6
Agent-Based Simulation on Residents' Travel Mode Choice for Local Transportation Development Strategy

6.1 Introduction

In last chapter we have introduced an ABM, namely ShopSim to support Japanese local government decision-making on the local large-scale shopping development strategy. In that work we combined a transportation choice module into the simulation of household agents' shopping choices. In that case the travel mode choices of household agents just be influenced by a factor named cost, other factors such as weather, social characteristics and household agents, travel destination, etc., have not been mentioned. But in a planning practice of transportation development management strategy, for example the improvement of public transit mode share, the influencing factors of individual travel mode choice are much more than the single factor of costs. Thus, in this chapter we will further introduce the travel mode choice model.

In recent years, with the acceleration of urbanization, the rapid development of urban transportation brings increasingly serious ecological environmental pollution, as well as energy supply constraints (Wu and Zhang 2018). In order to solve this problem, the government puts forward the strategy of building "bus city", which makes the prediction of bus sharing rate, as one of the important evaluation indicators, more and more attention is paid to. Therefore, it is urgent to develop four green and low-carbon ways of transportation. It is mentioned in the 13th five-year plan of urban public transport that "we will thoroughly implement the public transport priority development strategy, take public transportation development as the starting point, and build an efficient, convenient, safe, comfortable, reliable, green and low-carbon urban public transport system to meet the basic travel needs of the people". As the primary content of the "transit metropolis" assessment system, the public transit mode share has always been an essential index for cities formulating traffic-related planning strategies.

Y. Ma, Z. Shen, *Strategic Spatial Planning Support System for Sustainable Development*, Advances in Geographic Information Science,
https://doi.org/10.1007/978-3-031-07543-8_6

But facing to the planning strategy, what kind of policy solutions will positively effect on the improvement of the public transit mode share? This is still a research issue. Regarding to it we proposed an agent-based Resident Bus (RB) model to describe the mechanism of residents' transportation choices, predict the public transit mode share, and evaluate the implementation effects of different planning policies. According to the Transportation Development Management (TDM) strategy, three different planning scenarios are proposed in this work. Through evaluating on the implementation effects of each scenario, the results are supposed to support local government and planner's decision-making on relevant planning policies.

How to improve the public transit mode share through planning is a problem that needs to be solved by planners and has been concerned by scholars from all over the world. Existing studies on public transit mode share mainly include the choice of travel mode and influencing factors, which can be summarized as calculation, evaluation and prediction, among which the corresponding prediction methods can be divided into two categories: aggregate model and disaggregate model. Aggregate model is to conduct statistical treatment and analysis of each person's traffic activities according to the traffic area. The disaggregate model gives the concepts of maximum utility and random utility, takes the individual as the analysis object, and directly uses the individual information to construct the model which can describe the travel decision-making process of individuals or families (Chen et al. 2010).

The aggregate model is based on the statistics and calculation of the traffic zone, but does not describe the individual behavior. Therefore, it is often difficult to describe the individual's travel choice behavior, and the relevant research only stays on the surface of the macro phenomenon that resulted from individual behaviors (Ma et al. 2019). In order to improve the research limit by using aggregate model, the disaggregate model was gradually developed in the 1960s. The disaggregate model follows the logic of random effectiveness, and reflects the traffic mode choosing behavior through the description of traveler characteristics. Among them, some scholars use logit model to build a prediction/analysis model by describing the travel decision process of individuals or families (Liu et al. 2008; Bhatta and Larsen 2011). Such studies can better describe the attractiveness of traffic modes, but individual's travel behavior is usually connected to given circumstances, for example the characteristics of individual, his/her commuting preference, transportation facilities, ect. So some scholars manage to build multinomial logit model or mixed logit model, in order to reflect the complex mechanism between travel patterns, commuter's individual characteristics, family characteristics, seasonal changes travel mode choice (Li and Xu 2015; Thomas et al. 2019). In general, the disaggregate model has been constantly developed and improved, and its adaptability has been constantly improved. Moreover, it has gradually developed from simple Logit model to multiple Logit models and mixed Logit models, which are commonly used by researchers.

In summary, starting from individuals, the disaggregate model can more accurately describe the travel decision making process of individuals or families and calculate the public transit mode share. However, as a complex interconnected system, the behavior of individuals in a city is often not stable. On the contrary, it is often sensitive to the influences of external factors, such as the impact of weather changes on residents' traffic modes (Cui and Luo 2014; Zhang et al. 2018), residents' income on their traffic modes (Zhang et al., 2008), or the service quality of public transportation and so on. Although the existing researches have described and explained the relationship between individual behavior, urban traffic system and public transit mode share in a deep and comprehensive way, but still lack of description of the complexity and uncertainty of individual travel behavior, especially the mechanism between the interactive process of "planning intervention-personal behavior feedback-macro pattern of public transit mode share ". This is a key point that significantly related to the prediction on the outcomes of implementation effect of relevant planning policies. Thus, researching on the above mechanism is closely connected to how we decide on the policies regarding to improve the public transit mode share. Planning decisions. It is exactly the focus and purpose of this study.

As a model reflecting the interaction between individual behavior and environment, as well as the interaction between individuals, the agent-based model can effectively describe the interactive process between human and environment and the uncertainty of individual decisions in a complex system (Batty, 2005a, b). This method is widely applied in the research field of sociology, its application in urban planning first appeared in 1970s. Schelling proposed the first sociological ABM in 1971 to explain racial segregation in American cities (Schelling, 1971), Since then, it has been widely applied in related fields, and along this direction, it has gradually started to consider various phenomena in cities, including geographical space and social, economic, ecological environment, transportation, ect (Liang, 2012). Some scholars studied urban growth by developing Agent-based models for real estate developers and residents (Kaviari et al., 2019), while others studied the interaction between land use and transportation (Motieyan and Mesgari 2018).However, existing researches mostly use Agent-based model to analyze and predict individual behavior under specific conditions (Gurram et al. 2019; Kaviari et al. 2019), or the analysis of individual behavior in order to reflect environmental impacts (Manley and Cheng 2018; Fu and Hao 2018), there are few studies on "planning intervention – personal behavioral feedback". Although some scholars have developed ABMs to support the decision-making support of planning policies in the field of urban planning (Ma et al. 2013, 2016, 2019), but not yet involved in the impact analysis of public transit mode share.

In general, in the field of urban planning, although some scholars have paid attention to the development of ABM for planning decision support, in the field of urban transportation, they mainly focus on the research on the spatial scope and traffic flow (Melnikov et al. 2016; Manley and Cheng 2018), pays less attention to the urban public transit mode share, and there are few studies that take the travel choice model and ABM as a coupled model to carry out the quantitative analysis on the impact of planning policies by changing planning strategies. This chapter aims to portray the mechanism of residents' travel mode choice following the logic of "planning intervention – personal behavior feedback – macro pattern of public transit mode share ".

The remainder of this paper is organized as follows: The second part is to introduce the structure of the model. The third part is the application of the model, in this part, the parameter setting of the model is described, and the model is tested and the scenario is simulated. Finally, the research results are summarized and discussed.

6.2 Methodology of TDM Decision-Making Support

6.2.1 Planning and Decision-Making Process

This paper uses the method of dynamic simulation to build an ABM involving residents, public transport and planners, analyzes the improvement of public transit mode share under different planning strategies, and seeks for practical and effective schemes, so as to provide decision support for urban planning policies on improvement of public transit mode share.

As we introduced in Chap. 1, when an ABM developed for decision-making support of targeted planning policies, the planning and decision-making process should be developed and combined into the ABM model. in this case, a planning and decision-making process of RB model is designed as shown as Fig. 6.1, which is mainly composed of model simulation and planning decision-making. In terms of model simulation, this model firstly set up simulation related geographic data, bus information data and weather data as the preliminary dataset of the model, and then takes the personal information data obtained through questionnaire survey as the input conditions of the model, and then conduct the model simulation in order to output the simulation results under a center input. Due to the input is designed to reflect a special planning conditions, the relevant outputs are supposed to be able to reflect the implementation feedbacks of the policy. Among them, the geographical data mainly refers to the basic environment of bus station and road network in a case study area. When residents choose a travel mode, this individual choice will be affected by their own age, family characteristics and other personal social status, as well as the spatial distance between the destination and origin, the statues of public transportation, such as accessibility of bus stops, departure interval of bus and the service quality. The operation flow chart of RB is showed by Fig. 6.2.

In the aspect of decision-making of planning policy, the optimal planning decision is made by simulating the situation according to the change of related parameters of different strategies. The planning strategy used in this work mainly involves TDM strategy. TDM, stands for a strategy that targets to reduce the share of individual vehicles in the transportation network, improve the attractiveness of public transportations (such as bus) through a pull-to-push strategy. Compared with the traditional principle of cubic convergence, which is usually used to control the supply side, tends to add new transportation infrastructure, such as roads, highways, Bridges and underground passages (Kornhauser and Fehlig 2003), TDM strategy is less limited by conditions such as space and financial investments and shows higher applicability. Pull strategies motivate or attract travelers to use sustainable modes of transportation and push strategies to prevent them from using unsustainable modes of transportation. Generally, the pull strategy is balanced with the push strategy to achieve an effective travel demand management solution and improve the public transit mode share to ease congestion (Machemehl et al. 2013).

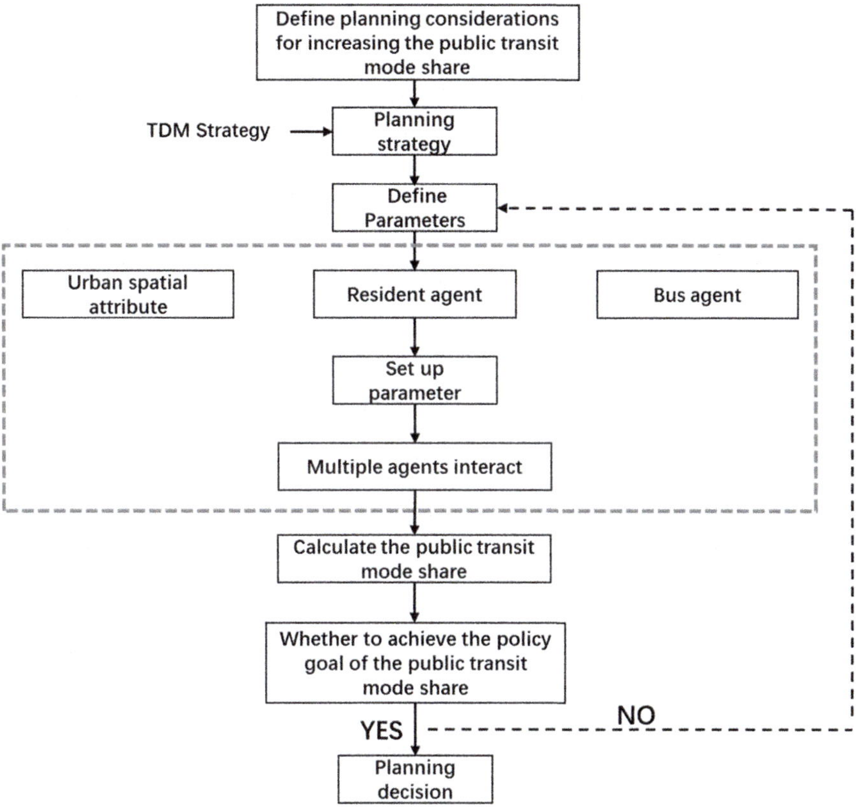

Fig. 6.1 Planning and decision flow chart of public transit mode share based on ABM

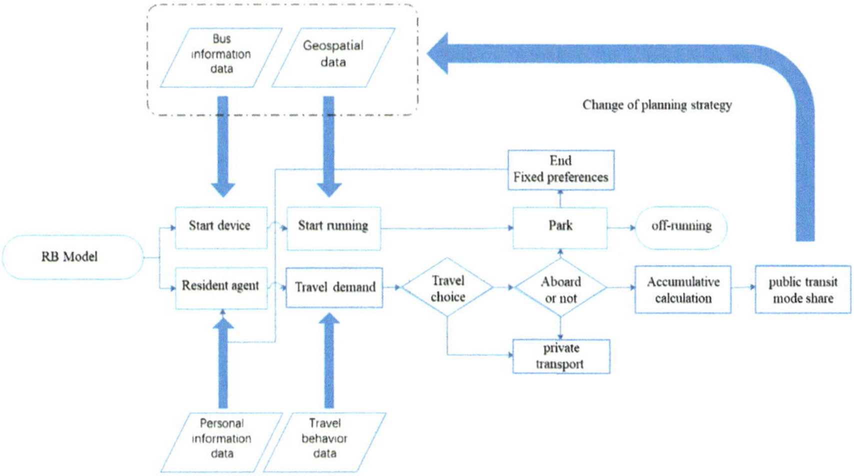

Fig. 6.2 Operation flow chart of RB model

As for the application of model in planning decision support, this study aims to verify the effectiveness of TDM strategy, and upgrade the TDM strategy from "qualitative" to "quantitative" by changing the input parameters and initial conditions of ABM model, so as to explore the optimal TDM planning strategy, and improve the public transit mode share.

6.2.2 The Framework of RB Model

Residents' individual attributes and behavior characteristics, bus attributes and weather conditions are the main factors affecting whether people choose public transportation. The RB model designed in this chapter is divided into three parts: input layer, influence layer and output layer. The input data is composed of four parts: individual behavior data, travel behavior data, geospatial data, and bus information data which describes the property, travel behavior, geographic space and inner space of the bus in the model. The influence layer is the input data that affects the residents and the bus behavior. The interaction between the bus behavior and the resident agent generates the preferences of different residents on the choice of public transportation. Finally, the result of travel choice is obtained through the output layer, and the public transit mode share is calculated. The framework of this study is showed as Fig. 6.3.

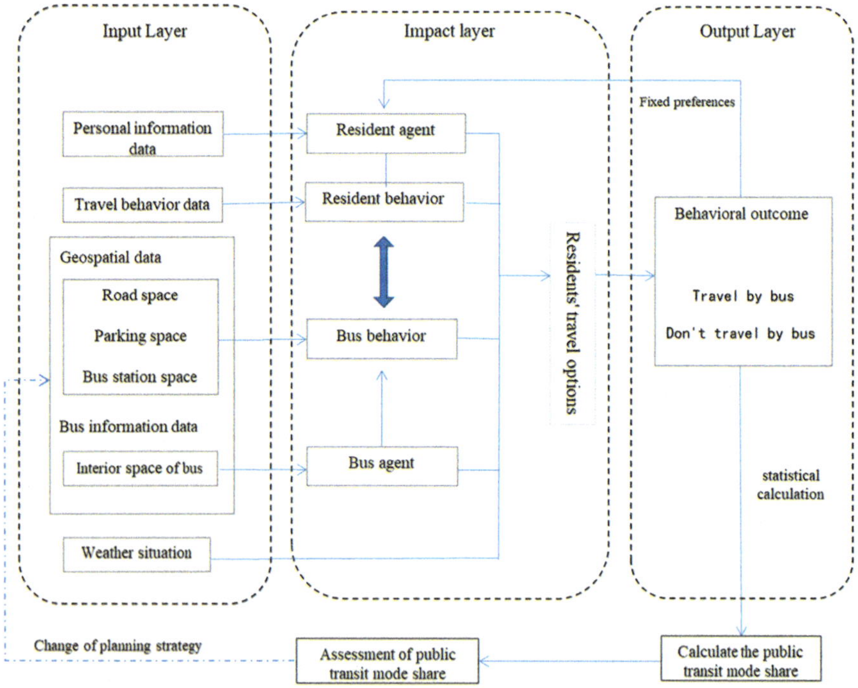

Fig. 6.3 The research framework of agent-based simulation on TDM decision-making support

6.3 Model Formulation

6.3.1 *The Environment of RB Model*

The environment of this simulation is different from the models we introduced in Chaps. 4 and 5, in which the environment is designed as a virtual urban space basing on local City Planning Law and spatial constraint regulations. This work is focusing on the bus operation and relevant transportation choices of residents. Thus, the environment should be able to accommodating the residents' distribution, bus stations and the routes of buses. In this case, the environment of RB model should include the road network of the research area. Figure 6.4 shows the geographical information of the environment we utilized in this study. It is the real geographical data of research area based on the real geographic space of baidu data map in Fuzhou City, China.

6.3.2 *Resident Agent*

In this work, resident agent is defined as an independent person, who has a residence and needs to commute to work, he/she can also make decision on commuting mode choice independently. The resident agents have gender, age, occupation, education, car ownership, ownership of motorbike, ownership of motorbike, commuting mode preference, entertainment travel preference, the basic properties of geographical location (longitude, latitude). All these attributes are related to the choice of travel mode, and their behavior rules are shown in Fig. 6.5.

According to the basic attributes, the resident agent generates the travel demand and decides whether to choose the bus or use the private transportation. Residents will first evaluate on the utility of different transportation modes. This process produces an original preference of resident agent for different travel mode choice. Then, who choose to travel by bus arrive at the bus station and wait for the bus. If the waiting time exceed the tolerance of waiting for bus, they will face to a choice to take private transportation. This situation also happens when they find the bus is full. Finally, the choice of the resident agent will be made according to the results of bus waiting. This process is showed by Fig. 6.5.

In the model this decision-making process of resident agents for different transportation modes is provided by a module of residents' travel choice. In this module, the logit model is used to evaluate the probability of commuter and entertainment travel mode choices. The model of commuting and entertainment travel is showed by Eq. (6.1):

$$\text{logit}(\text{P}) = \alpha + \beta_1 X_1 + \beta_2 X_2 + \beta_3 X_3 + \varepsilon \tag{6.1}$$

Fig. 6.4 The environment of RB model

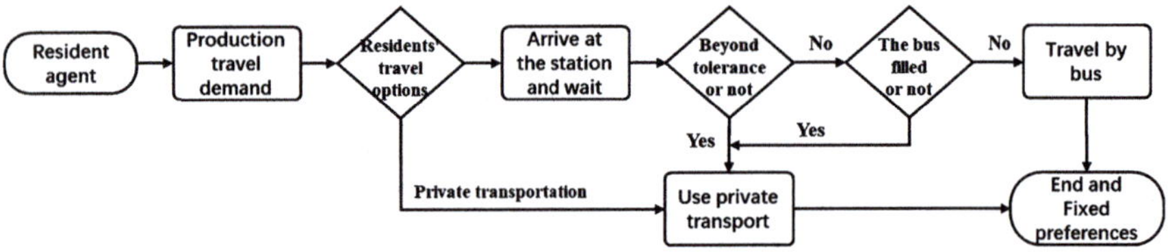

Fig. 6.5 Behavior rules of resident agents

Where

α is the intercept term and

β is the regression coefficient. If the value of β is positive, it means that with the increase of explanatory variables, the probability of that resident agents choose to travel by bus will increase

X_1 is the travel behavior variable

X_2 is the bus information variable

X_3 is the service quality variable of bus, and

ε is a random term.

6.3.3 Bus Agent

In this work, the bus agent is defined as a city bus in operation with a certain passenger carrying capacity. The bus agent has the basic attributes such as bus number, departure interval, schedule for the first and last bus, origin bus station, destination, route, bus stations on the route, distance from the next bus station, speed, bus comfortable degree and crowded degree. The decision rule of the bus agent is shown in Fig. 6.6. Start after each bus agent reaches the departure time, if there is no traffic jam, drive normally and stop at the station. If the traffic jam is encountered, slow down the driving; if the traffic jam is alleviated, normal driving will be recovered. Otherwise, continue to slow down the driving until the station is reached. If the bus agent arrived at the terminal stop, the driving will be ended. This simulation loop will be repeated till the bus agent arrived at the terminal.

6.3.4 Agents Interaction

For the city, which is a complex adaptive system, the interaction between its elements leads to the change and development of the city, so the interaction between agents is very important. The bus agent and the resident agent have their own behavior rules, but they also interact and influence on each other. The interaction between the two types of agent is shown in Fig. 6.7.

The model of this study is simplified in proportion to the population distribution of a city, and 1000 resident agents are generated to distribute in the model. The model only considers the condition of road congestion or traffic status. Due to the lack of traffic light data in a city we do not input the traffic light into the model but we assume that there will be a traffic light in each traffic congestion and all the bus agent will follow the constraint of traffic light very well. In this case, we ignore the traffic light's effects on driving speed and the time cost. Meanwhile, we do not consider the effects caused by traffic accidents on bus traveling. The model does not consider the expansion of urban space and growth of urban population. The model only simulates the commuting from 0 to 12o'clock in the morning of working days and the entertainment travel during holidays. Moreover, it is assumed that residents

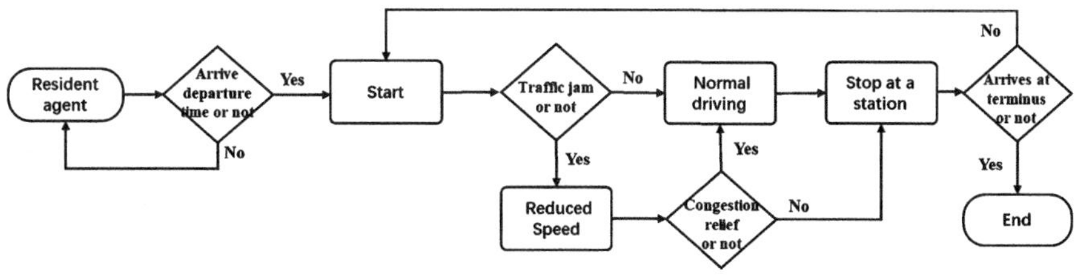

Fig. 6.6 Decision rules of bus agents

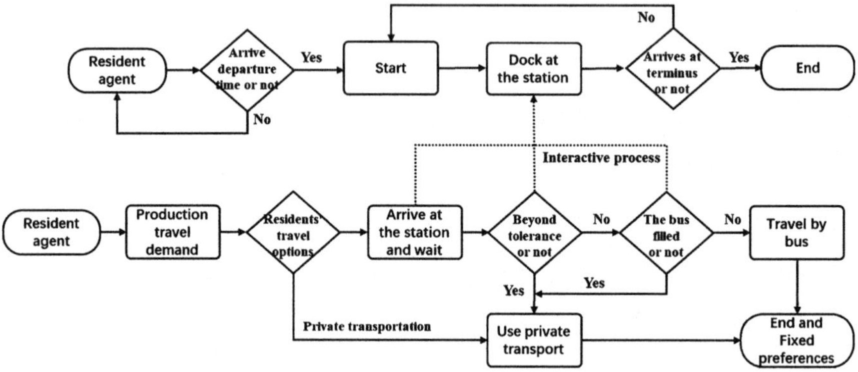

Fig. 6.7 The diagram of agent interactions within RB model

who have jobs must work from Monday to Friday, generate entertainment travel demands on Saturday and Sunday, and do not generate work demand on that date.

6.4 Variables and Parameters for Model Application

6.4.1 Case Study Area

According to the density analysis chart of bus stations in Fuzhou City, China (Fig. 6.8), the bus stations are mainly concentrated in the city center area. The spatial pattern of bus station in the urban space shows a density decrease from the city center area (Gulou district and Taijiang district) to the suburban area (Cangshan district, Jin an district and Minhou district). Within this analysis, we classified the bus stations into ordinary bus stops and major bus stations, the latter stands for the top 35 stations with most bus line ownership. The density of major stations is showed by Fig. 6.9.

6.4.2 Variable Description

1. Explained variable: bus travel choice

The bus travel choice of residents is the explanatory variable of the travel choice model for resident agents. The variable is further divided into the variables for commuting bus choice and entertainment bus choice. In view of the stability of travel choice in the medium and short term, the recent commuting bus choice and the entertainment bus choice of the sample shall prevail.

2. Explanatory variable: travel behavior data, bus information data

According to the travel behavior data of agents, it can be divided into the way they go to a bus station within the threshold distance, the time to go to the bus station, the waiting time at the bus station, whether to change the bus, the time to change

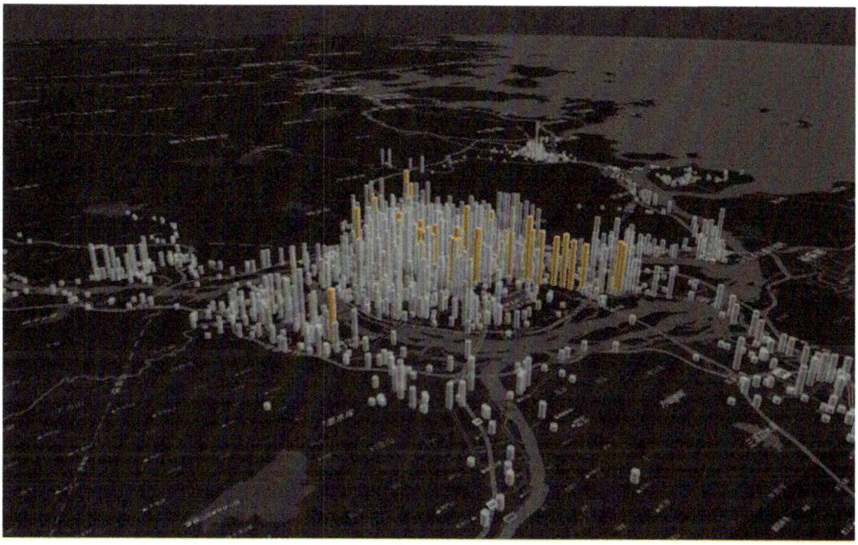

Fig. 6.8 The spatial distribution of bus stations in Fuzhou City, China

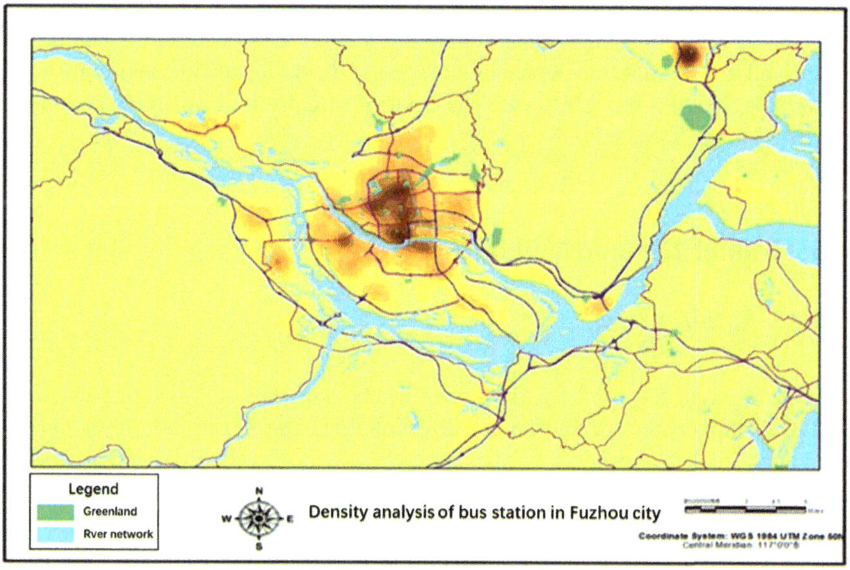

Fig. 6.9 The spatial distribution density of the major bus stations in Fuzhou city, China

the bus, the time inside the bus, and the time from the destination station to the work place. At the same time, resident agents also have a variable of bus waiting tolerance, which is used to describe the longest bus waiting time basing on questionnaire. And the comfortable degree of the bus is also took as an explanatory variable. The comfortable degree includes the crowed degree of the bus and the environment in the bus.

3. Confounding variable: personal information data, weather data

The residents' age, gender, education level, occupation, family income and transportation ownership also affect the choice of travel mode, so these variables need to be controlled. Weather condition is also an important factor that restricts the travel mode choices of resident agents. This study controlled this variable by the method of questionnaire scenario design.

The detailed variables and settings are showed by following Table 6.1:

6.4.3 Initial Conditions for the Simulation

During the creation and operation of the model, a total of four types of data were input: real geospatial data and bus information data of Fuzhou city, individual information data, and weather data. The data of individual behavior and travel behavior mainly come from questionnaire survey.

The parameters of the simulation are divided into global parameters and agent related parameters. Regarding to the global variable, it contains the bus station location, the proportion of city residents in different districts, the weather of a way, the time for first and last bus in a day, working day or holiday, and the environment of the simulation. The agents' data include the social status of residents, the driving statue and the comfortable degree of bus agents (including ainside environment and crowed degree of a bus gent), ect. The detailed initial settings of these parameters are listed in the Table 6.2.

6.5 Model Test and Validation

6.5.1 Stability Test

The agent-based RB model is developed by the software of Anylogic, which is a software supporting for multimethod simulation. Users can use it to develop dynamic simulation models by the methodologies of system dynamics, discrete event modeling, and agent-based modeling or in any of the combination of these three methods. In AnyLogic, you can use various visual modeling languages: process flowcharts, statecharts, action charts, and stock & flow diagrams. In this work we used the language of process flowcharts to develop the RB model, and a screenshot of the model running is showed by Fig. 6.10.

In order to test whether the public transit mode share simulation is stable, the model was repeated for 10 times and each time 14 loops with the above initial parameters unchanged. Each simulation loop represents 1 day in real life, 14 loops represent 2 weeks. During the test, we repeated the experiments for 14 times and put the outputs into statistical analysis. The differentiations of the simulated public

Table 6.1 The variables of the simulation

Variable	Explanation
1, Explained variable: bus travel choice	
Commuting bus travel options	Yes is 1, no is 0
Entertainment bus travel options	Yes is 1, no is 0
2, Explanatory variables: travel behavior data, bus information data	
Waiting tolerance	The longest bus waiting time tolerated by the respondents, unit min
Time to get to the bus station	Time of respondents to the bus station, unit min
The way to the bus station(walk)	Yes is 1, no is 0
The way to the bus station(cycle)	Yes is 1, no is 0
The way to the bus station(motorbike)	Yes is 1, no is 0
The degree of transfer demand	Most need: 1, occasionally need: 2, do not need: 3
Time cost for transfer	Additional time required by respondents to transfer, unit min
Tolerance for bus waiting	Waiting time at the bus station of the respondents, unit min
Time cost from bus station to destination	Time required for respondents to travel from the destination site to the workplace, unit min
The time in the bus	The time required by the respondents in the bus body, unit min
Comfortable degree of the bus	There are four grades: 1, 2, 3 and 4. 4 is the most comfortable
3, Confounding variables: personal information data, weather data	
Age	Age of respondents
Gender	Gender of respondents, male: 1, female: 0
Education level	There are four grades: 1, 2, 3 and 4. 4 is the highest
Occupation(student)	Yes is 1, no is 0
Occupation(Enterprise staff)	Yes is 1, no is 0
Occupation(project units)	Yes is 1, no is 0
Occupation(privately-owned)	Yes is 1, no is 0
Occupation(retirement)	Yes is 1, no is 0
Family income	There are four grades: 1, 2, 3 and 4. 4 is the highest
Private car ownership	Yes is 1, no is 0
Motorbike ownership	Yes is 1, no is 0
Bicycle ownership	Yes is 1, no is 0
Weather conditions(hot)	Yes is 1, no is 0
Weather conditions(sunny)	Yes is 1, no is 0
Weather conditions(rainy)	Yes is 1, no is 0

transit mode share within each experiment is showed by Figs. 6.11, 6.12. The former represents the stability of the model in simulating the public transit mode share of commuting travel and the Fig. 6.11 represents the stability of the model in simulating the public transit mode share of entertainment travel. The statistical results show that the average sharing rate of the RB model for commuting travel fluctuates between 0.6 and 0.63, while that entertainment travel purposes fluctuates between

Table 6.2 The initial parameters and settings of RB simulation

	Parameter name	Units	Initial value
Global variable	Bus station location	(x, y)	*
	Simplify the urban population	Num	10,000
	Bus station environment		2.5
	Weather conditions		Sunny
	Working days/ holidays		Working days
	Running start time	Min	360 (note:6:00)
	Running end time	Min	720(note:12:00)
Bus agent attribute	Bus no.	Int	*
	Departure interval	Min	*
	First and last bus time	Hour	*
	The station to get on	(x,y)	*
	The station to get off	(x,y)	*
	The distance to the next bus station	km	*
	Speed	km/h	40
	The comfort degree of the bus		2
	The number of passengers	Num	0
Resident agent attribute	Age		[9,60] generated by probability*
	Gender		Variate{0,1} generated by probability*
	Start position	(x,y)	*
	Destination	(x,y)	*
	Education level		Generated by the probability distribution of the questionnaire*
	Occupation		Generated by the probability distribution of the questionnaire*
	Family income	Chinese Yuan	Generated by the probability distribution of the questionnaire*
	Family owned vehicle		Generated by the probability distribution of the questionnaire*
	Tolerance for commuting	Min	Generated by the probability distribution of the questionnaire*
	Tolerance for entertainment travel	Min	Generated by the probability distribution of the questionnaire*
	Transfer or not		Generated by the baidu map API*
	Time to get to the bus stop	Min	Generated by the baidu map API*
	The way to the bus station		Generated by the probability distribution of the questionnaire*
	Waiting time(commuting)	Min	Generated by the baidu map API*
	Waiting time(entertainment)	Min	Generated by the baidu map API*
	Bus station to workplace time	Min	Generated by the baidu map API*
	The time in the bus	Min	Generated by the baidu map API*
	Time cost for transfer	Min	0: No need to transfer [5,20] transfer*

*Refers to the parameters which values are input from real dataset iaccording to the situation of case study area

0.45 and 0.48. Its standard deviation is between 0.02 and 0.06, which is an acceptable range.

6.5.2 *Sensitivity Test*

1. Change in the variable of waiting tolerance of resident agents

In the sensitivity test, the experiment will be repeated 20 times according to the initial parameters, each time representing 14 days in real life. The simulation results of commuting travel fluctuate around 61%, while that for entertainment trips fluctuates around 46% (as showed by Figs. 6.13, 6.14, respectively). After re-initialization, the waiting tolerance was increased by 50% with other parameters unchanged, and the relationship between public transit mode share and waiting tolerance was obtained.

In the two experiments, the simulation results are relatively stable and the results show the changes in public transit mode share resulted from the adjustment waiting tolerance of resident agents. The results of the public transit mode share for the commuting travel fluctuates around 65% after the adjustment of waiting tolerance improved in the input. And for the entertainment travel, the public transit mode share fluctuates around 50% after the tolerance improved. The sensitivity test proves the validity of this parameter to the public transit mode share.

2. Change in the variable of bus driving speed

Similarly, in this test experiment is repeated for 20 times and each experiment stands for 14 days in real life. We adjust the bus driving speed from 30 km/h to 50 km/h, as showed by Fig. 6.15, the simulation results of public transit mode share increased from 28.12% to 41.29% correspondingly. The same method could be employed to test the sensitivity of other parameters in turn, such as weather, bus station environment, departure interval. In this case we just chose the representative parameters that related to agents' behaviors to do the sensitivity test, other global parameters have not been introduced here.

6.5.3 *Validation by Real Data*

Before validating the RB model, the parameters should be set according to the real situation of case study area. In order to set the parameters, we investigated the utility preferences of resident agents for travel mode choices by a questionnaire survey in Fuzhou City, China. During April 4th, 2018 to April 19th, 2019, we conducted a questionnaire investigation of travel mode choice in Fuzhou City, China. The investigation was carried out in the center city area of master urban, which covers 5 main districts of Fuzhou City. These districts are Gulou district, Taijiang district, Cangshan district, JinAn district and Mawei District. 600 copies of questionnaires

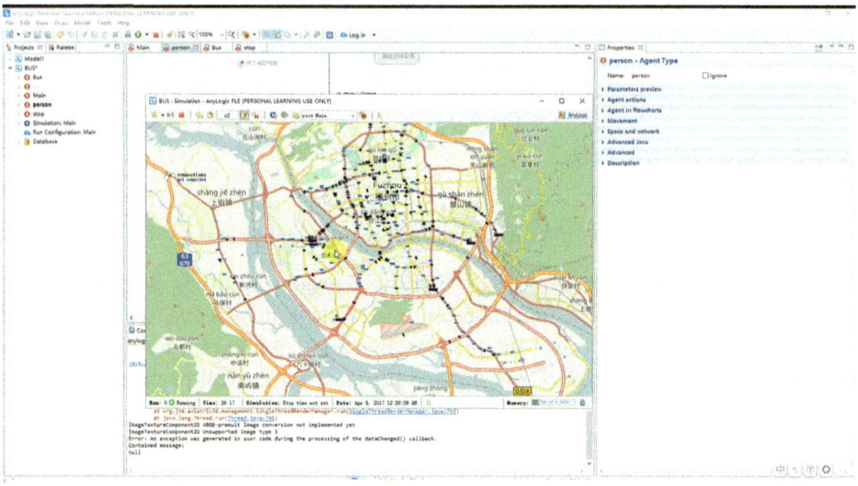

Fig. 6.10 The screenshot of RM model running

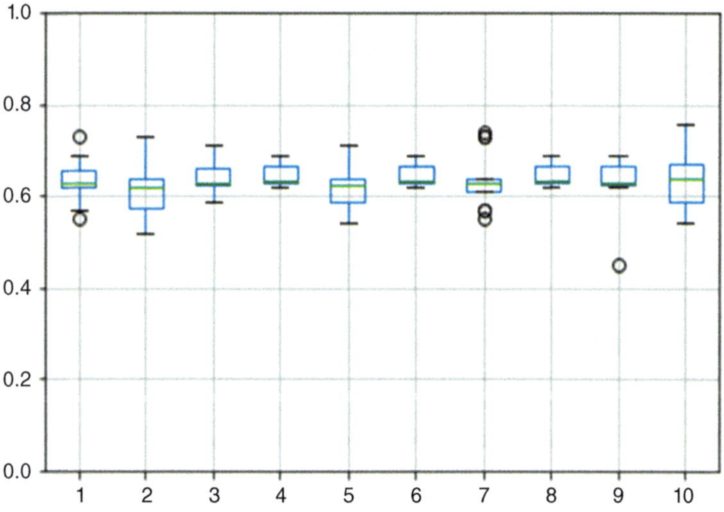

Fig. 6.11 The stability test of RB model for commuting travel

were sent and 585 respondents were collected in total. Within the questionnaire, there are 38 questions regarding to individual information, commuting travel mode choice and entertainment travel mode choice. All the parameters for the utility model of travel mode choices were included in the questions. In the meantime, at the end of the questionnaire we also asked them about their travel mode in last 14 days respectively to commuting and entertainment travel. Basing on the questionnaire survey we conducted partial correlation regression analysis and received

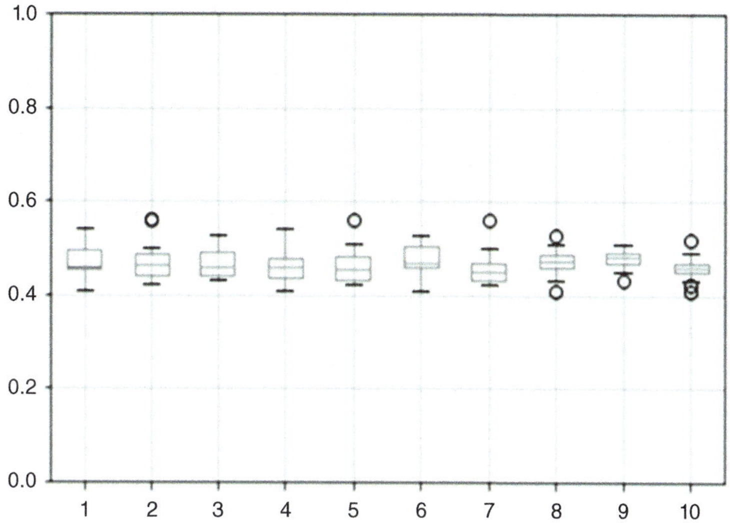

Fig. 6.12 Stability test of RB model for entertainment travel

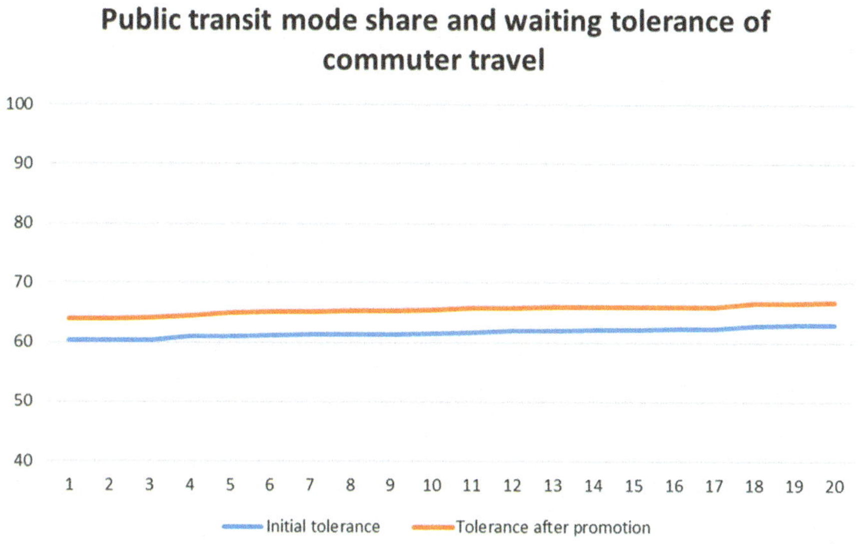

Fig. 6.13 The sensitivity test of the variable of waiting tolerance for the commuting travel

the coefficients for the variables that would affect residents travel mode choices. The results are showed by following Table 6.3:

In order to verify whether the simulation results of the RB model are close to the real situation of the case study area, it is necessary to compare the simulated public transit mode share by the RB model with the real public transit mode share data of Fuzhou City. The reference data selected for verification is the real public transit

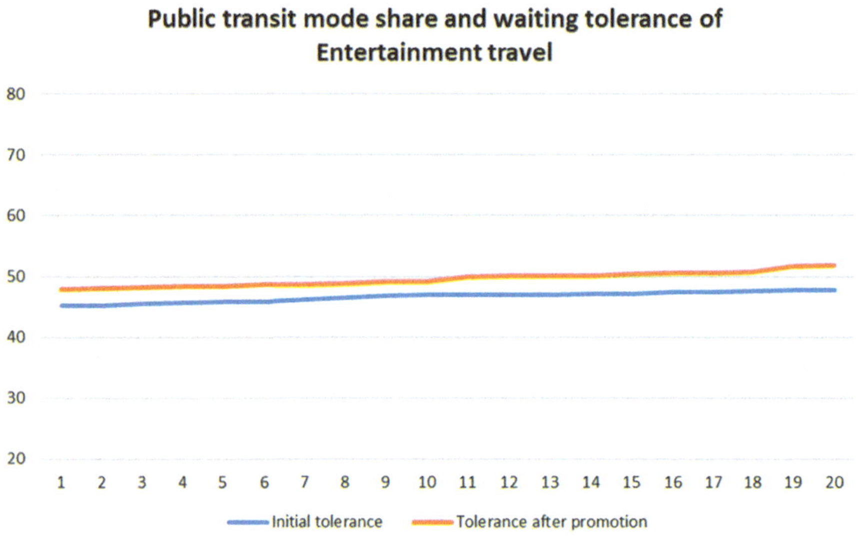

Fig. 6.14 The sensitivity test of the variable of waiting tolerance for the entertainment travel

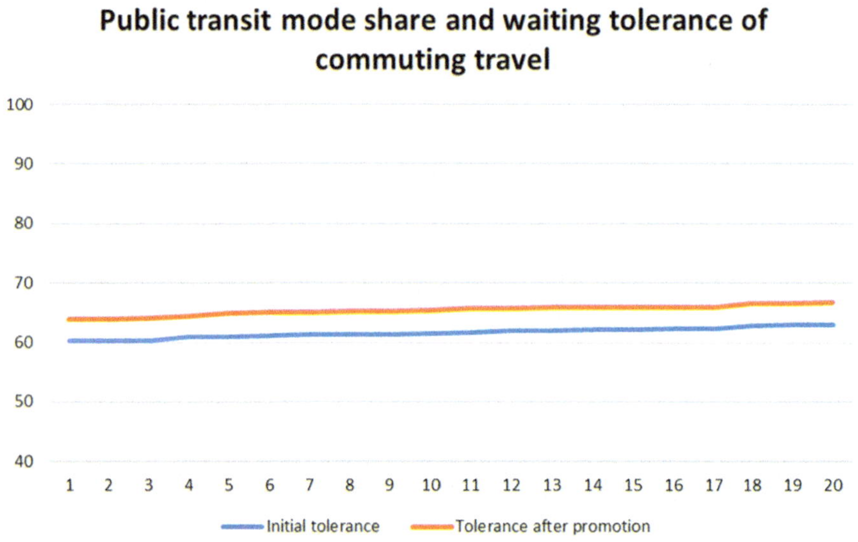

Fig. 6.15 The sensitivity test of the variable of bus driving speed for the commuting travel

mode share which is 46.5% of Fuzhou in the March 2018 (referenced to the link: http://fzjt.fuzhou.gov.cn/zz/zwgk/tzgg_33472/201909/t20190906_3032884.htm). Due to this data has not been reveal by local official statistic data, the figure we used here is sourced from website regarding to a news of local government conference. Meanwhile, in the news the public transit mode share is not separated from

Table 6.3 Partial correlation coefficient of impact factors on resident travel mode utility

Variables	Coefficients for commuting travel	Coefficients for entertainment travel
The way to the bus station(walk)	0.352***	0.814**
The way to the bus station(cycle)	1.006***	1.387***
The way to the bus station(motorbike)	−0.352	−0.421
Occupation(student)	−0.442	0.859**
Occupation(Enterprise staff)	0.226*	0.009
Occupation(project units)	1.029**	−0.201*
Occupation(privately-owned)	−0.642*	−0.624
Occupation(retirement)	−0.197	−0.533
Age	0.002	0.027*
Gender	0.071	0.153
Education level	0.155*	0.250**
Family income	−0.072	0.007
Private car ownership	0.110	−0.390***
Motorbike ownership	0.117	0.296
Bicycle ownership	0.207	0.071
Tolerance for bus waiting	0.012*	0.036**
Time to get to the bus station	−0.002	−0.008
The degree of transfer demand	0.459***	0.378**
Time cost for transfer	−0.041	−0.025
Tolerance for bus waiting	−0.026**	0.010
Time cost from bus station to destination	−0.007	0.015
The time in the bus	−0.014	0.008
Comfortable degree of the bus	−0.195*	−0.248*
Weather conditions(hot)	−0.064	−0.064
Weather conditions(sunny)	0.037	0.037
Weather conditions(rainy)	0.012	0.012

Significant values: 0.05 "*" 0.01 "**" 0.005 "***"

commuting travel and entertainment travel. In this case the accuracy of the data could not be guaranteed. Here we just use it to introduce the validation method in this model research and the validation with official statistic data will be the future work.

For validation we repeated the experiment for 20 times, each time stands for 14 days. The simulation results of public transit share have been output and average values about the share rate for commuting travel and entertainment travel are calculated. The average public transit share rate for commuting travel is between 60% and 63%, and that for entertainment travel is between 45% and 48%. The comparison shows that the simulation results are higher than the real data.

After analyzing the cause of this result, we concluded two possible reasons:

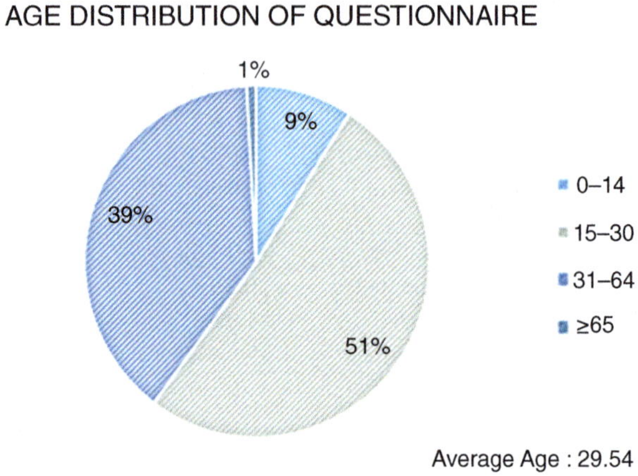

Fig. 6.16 the age distribution of respondents in the questionnaire survey

1. the respondents from residents who are younger than 30 years old took 60.18% of the total respondents and the average age of respondents is about 29.54 years old (as shown as Fig. 6.16). This group of residents themselves show high bus preferences in real life. In the questionnaire due to the residents elder than 35 years old are not very cooperate so that the ratio of this group respondents is too high. This is probably the main reason why the simulated public transit mode for commuting travel is more than 60%, nearly 10% higher than the data published by local news.
2. In the simulation, a variable namely the time cost to but stop, in the simulation is calculated by Euclidean distance between the location of resident agent and the nearest bus stop. While in real, the time cost will be influenced by the accessibility from resident location to bus stop, and the distance is basing on real road network. It means the simulated time cost probably shorter than real. While will influence the resident agents' utility evaluation. It is easy to be understood, in the questionnaire we also found that one of the key factors affects residents' bus choices is accessibility to bus stop. So that the shortened time cost to bus stop may lead to a higher public transit share.

However, in the questionnaire we also asked the travel mode of residents within last 14 days, basing on the responds we calculated the public transit share rate for commuting and entertainment travel. The figures are respectively 61.01% and 47.92%, which are near to the simulated results (the differentiation is within 2%).

6.6 Scenario Analysis

6.6.1 Scenario Design

In questionnaire survey, to verify the planning strategy and reveal the causes that lead residents not to choose public transportation mode. In terms of commuting travel, the reasons for residents who do not choose bus are:

1. the route is not direct;

2. the bus is crowded
3. the waiting time is long with few buses.

While for entertainment travel, the reasons for residents who do not choose bus are:

1. the route is not direct;
2. the bus is crowded.

In addition, regarding to a question of TDM suggestions, the questionnaire supplied seven candidate improvement policies for residents, and the choices of respondents are shown in Fig. 6.17:

From the Fig. 6.16 it could be seen that regarding to the TDM to improve the public transit share in Fuzhou city, residents (round 30 years old) focus on improving the bus service level, increasing the number of buses, optimizing the bus routes and strengthening the management of bike-sharing. Therefore, based on the realistic background of local public transportation management and the residents concern in questionnaire feedback, this study proposes three planning policy scenarios based on the TDM strategy (Table 6.4):

6.6.2 *Strategy 1: Bike-Sharing Area Control*

In recent years, the popularity of bike-sharing has made it easier for people traveling within short distances, especially as a transfer tool for other transportation mode. Bike-sharing is a bike rental service provided by enterprises in urban areas. However, due to the lack of control, there is a very serious phenomenon of random parking

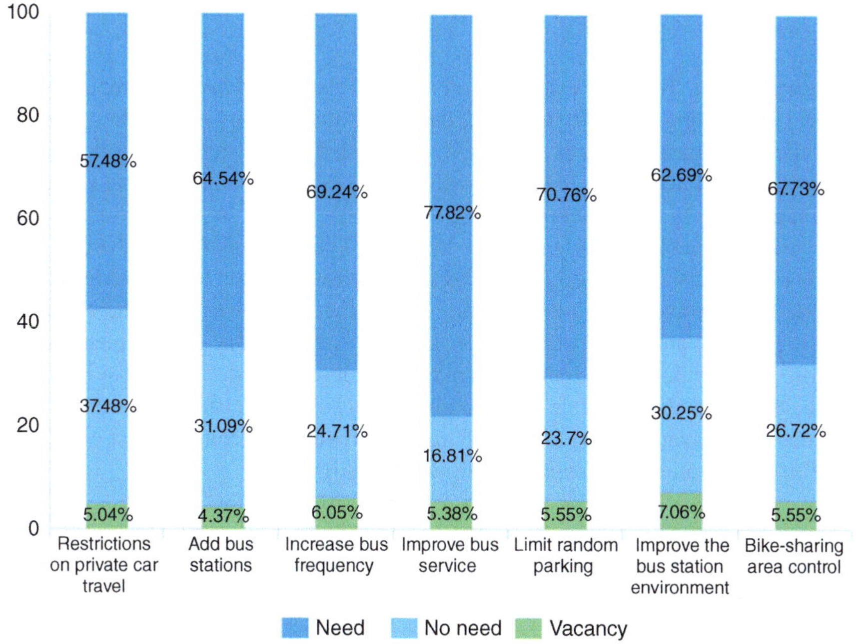

Fig. 6.17 Statistical results of respondents' concerns on TDM

Table 6.4 TDM strategy for improving public transit share

ID	Strategy	Category	Management direction	Travel variables that may be affected
1	Bike-sharing area control	Push	Road space	Change the way individuals travel to bus stops
				Reduce personal time to bus stops
2	Improve bus service	Pull	Bus interior space	Improve comfort from buses
3	Improve the bus station environment	Pull	Bus station space	Increases personal waiting tolerance

Table 6.5 simulation results of control bike-sharing area for commuting travel

Probability	0	0.1	0.2	0.5	1
Public transit mode share rate	63.71	65.07	66.71	70.84	77.21
The first quartile (25%)	62.01	63.25	64.31	69.02	74.52
The second quartile (50%)	63.05	65.54	66.86	70.47	77.93
The third quartile (75%)	64.78	67.71	69.12	72.79	79.73
The fourth quartile (100%)	67.04	74.89	71.04	79.96	81.95

Table 6.6 simulation results of control bike-sharing area for entertainment travel

Probability	0	0.1	0.2	0.5	1
Public transit mode share rate	47.10	51.07	52.28	58.94	64.35
The first quartile (25%)	45.25	49.04	51.79	56.83	60.00
The second quartile (50%)	46.10	50.16	52.52	57.77	63.14
The third quartile (75%)	49.73	53.41	54.08	58.72	64.93
The fourth quartile (100%)	50.90	56.02	56.21	62.73	65.12

and destruction. To some extent, the convenience of bike-sharing is reduced, and affects the urban road environment. Therefore, in this strategy design, it is assumed that the bike-sharing area is reasonably controlled, and when the resident agent chooses to walk to the bus station, it will choose to use the bike-sharing according to a random set in the utility model.

The influence of this strategy on the increase of public transit mode share in commuting travel is shown in Table 6.5, and the increase of public transit mode share in entertainment travel is shown in Table 6.6. In the tables we listed the simulated public transit share rate in second line, it is the average value of 20 times' experiments. Meanwhile, we output and ranked in a descend order. Then we listed the public transit mode share rate in the simulation results from of the experiments and calculated the average values of public transit share rates from the first quartile to the forth. The following Tables all followed this logic.

6.6.3 Strategy 2: Improve Bus Service

According to the results of the questionnaire survey, most of the opinions feed-backed from respondents are about increasing the number of bus lines. However, it is not within the scope of this paper to increase the number of buses and optimize the routes. This paper only considers improving the service level of public transport, and the comfortable degree of public transport. Thus, in this strategy we target to improve the service quality of bus lines with poor comfortable degree and environment. The bus lines targeted to be improved are listed in Table 6.7, which are the lines with bad reviews by respondent. As showed in this Table, the bad reviews of each bus line is different, in order to group them we use the breakpoint to divide them into 4 groups, the first group is the bus lines with more than 8 bad reviews (in the range of 8 to max), the second group is the bus lines with less than 8 but more than 5 bad reviews and the third group is the bus lines with less than 5 but more than 3 bad reviews, the last group is the bus lines with less than 3 bad reviews. The influence of the simulation results on the increase of public transit mode share for commuting and entertainment travel are showed through Tables 6.8, 6.9.

6.6.4 Strategy 3: Improve the Bus Station Environment

Due to many respondents also say that hope the bus environment could be improved. The environment of bus includes in car environment, the environment of bus stop. Improve the environment of the bus station, don't need to spend too much financial

Table 6.7 Bus line ID and the statistics of bad reviews in the questionnaire

Bus Line ID	2	55	20	1	22	11	9	3	117	75	8
Number of bad reviews	10	10	10	9	8	7	7	6	6	5	5
Bus line ID	66	129	5	100	167	29	52	6	18	87	37
Number of bad reviews	5	5	5	5	4	4	4	4	4	3	3
Bus line ID	301	125	113	501	27	317	128	23	77	85	13
Number of bad reviews	3	3	3	3	3	3	3	3	3	3	3

Table 6.8 simulation results of improve bus service level for commuting travel

Threshold	Max	8	5	3
Public transit mode share	63.71	65.32	67.34	69.95
The first quartile (25%)	62.01	64.86	64.27	66.31
The second quartile (50%)	63.05	65.33	65.36	70.02
The third quartile (75%)	64.78	66.62	68.79	71.38
The fourth quartile (100%)	67.04	69.29	72.22	73.36

support and it could also optimal the spatial utilization around the bus stop. Thus, it is designed as another strategy of this work. It is expected to improve the environment of the bus station by urban planning and design, provide convenient facilities, such as information service facility, sunshade and chairs at the bus station, so as to increase the tolerance of waiting time of residents for bus and improve the traveling experience of the residents. In this strategy design, it is assumed that in the case study area, the poor bus station environment is significantly improved, the waiting time of the resident agents who travels in the bus station increases d%, means that the maximum waiting time is increased by d%. The simulation results of public transit share for commuting travel is shown in Table 6.10, and the results for entertainment is shown in Table 6.11.

Through the simulation of the above three planning strategies, the public transit mode share can be improved to a certain extent by improving the level of a single item. Compared with the three strategies, strategy 2 has the most obvious effect and the highest improvement range. Strategy 1 has a certain effect and Strategy 3 has only a small increase in public transit mode share.

Table 6.9 Simulation results of improve bus service level for entertainment travel

Threshold	Max	8	5	3
Public transit mode share rate	47.10	50.32	54.49	56.84
The first quartile (25%)	45.25	48.68	52.73	53.81
The second quartile (50%)	46.10	50.18	54.56	56.39
The third quartile (75%)	49.73	51.26	55.25	59.37
The fourth quartile (100%)	53.90	55.94	58.92	62.16

Table 6.10 Simulation results of improve bus station environment for commuting travel

Increase tolerance by d%	0%	20%	50%
Public transit mode share rate	63.71	64.36	65.44
The first quartile (25%)	62.01	62.35	61.20
The second quartile (50%)	63.05	64.29	65.85
The third quartile (75%)	64.78	65.90	66.36
The fourth quartile (100%)	67.04	69.32	71.92

Table 6.11 simulation results of improve bus station environment for entertainment travel

Increase tolerance by d%	0%	20%	50%
Public transit mode share rate	47.10	47.57	49.18
The first quartile (25%)	45.25	45.42	47.66
The second quartile (50%)	46.10	47.39	49.25
The third quartile (75%)	49.73	48.20	50.73
The forth quartile (100%)	53.90	49.94	53.97

6.7 Discussions

This study aims at finding out the policy strategy for relieve urban traffic congestion and serious environmental pollution and practice the policy target of developing bus-oriented transit city through agent-based simulation. Basing on the utility theory, an agent-based RB model is developed to support local government decision makers making-decision on TDM for improving the public transit share rate. Regarding to a planning target but without very clear planning strategies, is the typical characteristic of research issue we are facing to in this study. It is different from the downtown revitalization through residential relocation allowance in Kanazawa City, the policy is already made and the research purpose is to investigate the possible effectiveness of the policy. In this work, we need to find out the possible policy solution to realize the target of improving public transit share rate. Thus, scenario is very important. This is similar with the case we introduced in Chap. 5. But the difference between the case introduced in last Chapter is that, in last chapter there are existing urban planning law and regulations on large-scale shopping center development. In this work there is no obvious policy references excepts for the planning constraint on bus stop by urban planning. But regarding to the TDM strategy, it much more focusing on management solutions, thus in this work we pay more attention on recognize the possible management scenarios.

In order to test the impact of different planning strategies on the public transit mode share, three strategies are designed basing on questionnaire in local city. Through the research of this work, it is a possible way to figure out the planning scenarios through questionnaire. But it requires a good design in the questionnaire and a well control of the samples. As the validation of real data, we introduced in this chapter, it could be seen that if the sample could not be well controlled, the coefficients would directly effect on the simulation results and hereby, make the results vary from real data. But due to the lack of official real data, it is hard to evaluate to what extent the simulation has been influenced in this work. While the effects must exist. Thus, the simulation results and the policy preferences all influenced by the majority of the respondents, who are around 30 years old.

From what has been simulated above, we may draw the three conclusion that: First, whether the residents need to change buses, the transfer mode, and the waiting time of residents at the bus station have a great influence on whether the residents choose to travel by bus; Second, through the control of bike-sharing areas to improve the bike-sharing service level, improve the bus service level can significantly improve the public transit mode share; Thirdly, improving the bus station environment shows the weakest effects on improving the public transit mode share. But these conclusions effected by the sampling in questionnaire, the accuracy needs to be improved. Meanwhile, in an ABM simulation, the parameter settings, especially the parameters for agents' decision-making are quite important and should strictly reflect the real situation. If these parameters are come from questionnaire, the sampling will influence on the simulation results.

References

Batty M. Cities and Complexity: Understanding Cities with Cellular Automata, Agent-Based Models, and Fractals. The MIT Press, 2005a.

Batty M. Cities and Complexity. Progress in Human Geography, 2005b, 31(1): pp. 113–115.

Bharat P. Bhatta, Odd I. Larsen. Errors in variables in multinomial choice modeling: A simulation study applied to a multinomial logit model of travel mode choice. Transport Policy, 2011,18(2): pp. 326–335.

Chen X, Zhou J, Li Z, et al. Evolutionary Analysis on the Dynamical Systems of Travel Behavioral Decision-Making[M]. IEEE, 2010.

Cui W.J., Luo Y. Effects of Urban Residents Climate Change Risk Perception on Travel Mode Choice: Based on 620 Urban Residents Investigation. Eco-economics, 2014, 30 (11): pp. 119–123+155. (in Chinese)

Manley, E., & Cheng, T. Exploring the role of spatial cognition in predicting urban traffic flow through agent-based modelling. Transportation Research Part A: Policy and Practice, 2018, 109: pp. 14–23.

F. Kaviari, M. Saadi Mesgari, E. Seidi, H. Motieyan. Simulation of urban growth using agent-based modeling and game theory with different temporal resolutions. Cities, 2019, 95: 102384.

Fu Z., Hao L. Agent-Based Modeling of China's Rural-Urban Migration and Social Network Structure. Physica A. 2018, 490: pp. 1061–75.

Gurram S., Stuart A.L., Pinjari A.R. Agent-based modeling to estimate exposures to urban air pollution from transportation: Exposure disparities and impacts of high-resolution data. Computers, Environment and Urban Systems. 2019, 75: pp. 22–34.

Kornhauser A.L., Fehlig, M.K. Marketable Permits for Peak Hour Congestion in New Jersey's Route 1 Corridor. TRB 2003, 2003. Washington DC: Transportation Research Board.

Liang C. Agent-based modeling in urban and architectural research: A brief literature review. Frontiers of Architectural Research, 2012, 1(2): pp. 166–177.

Li Q.H., Xu Y.N. Research on Commuter Travel Mode in Old City Based on Multinomial Logit Model. Modern Transportation Technology, 2015, 12(02): pp. 66–68. (in Chinese)

Ma Y., Shen Z.J., Mitsuhiko K. Agent-Based Simulation of Residential Promoting Policy Effects on Downtown Revitalization. The Journal of Artificial Societies and Social Simulation, 2013, 16(2).

Ma Y., Shen Z., Nguyen D.T. Agent-Based Simulation to Inform Planning Strategies for Welfare Facilities for the Elderly: Day Care Center Development in a Japanese City. Journal of Artificial Societies and Social Simulation. 2016, 19(4).

Ma Y., Wu R.H., Wang Z.Y., Li M.Y. Study on the Planning and Layout of Community Geriatrics Based on ABM: A Case Study of Downtown Fuzhou. Urban Development Research, 2019, 26 (01): pp. 18–25. (in Chinese)

Machemehl R., Kockelman K., Chen D., Bienkowski B., Smith L., ZhangZ. Travel Demand Management Guidebook. Texas: DoT&FHA.A, 2013.

Melnikov V. R., Krzhizhanovskaya V. V., Lees M. H., Boukhanovsky A. V. Data-driven Travel Demand Modelling and Agent-based Traffic Simulation in Amsterdam Urban Area. Procedia Computer Science, 2016, 80: pp. 2030–2041.

Motieyan H., Mesgari M.S. An Agent-Based Modeling approach for sustainable urban planning from land use and public transit perspectives. Cities. 2018, 81: pp. 91–100.

Schelling, T.C. Dynamic models of segregation. Journal of Mathematical Sociology, 1971, 1: pp. 143–186.

Thomas, T., Puello, L. L. P., & Geurs, K. Intrapersonal mode choice variation: Evidence from a four-week smartphone-based travel survey in the Netherlands. Journal of Transport Geography, 2019, 76: pp. 287–300.

Wu H., Zhang H.B. A brief analysis on the development and future of urban public transport in China. Electronic World, 2018 (17): pp. 12–15. (in Chinese)

Zhang F., Han Z.L., Chen Y.Y., Li J.X., Cui W. A Model of Highway Travelers' Travel Mode Shift under Rainy Weather. Highway Traffic Science and Technology, 2018, 35 (06): pp. 105–111+130. (in Chinese)

Zhang J.Y., Fujiwara Z.Z., Soe T. Capturing Travelers' Stated Mode Choice Preferences under the Influence of Income in Yangon City, Myanmar. Transportation System Engineering and Information, 2008 (04): pp. 49–62.

Chapter 7 Agent-Based Modeling for Decision-Making on the Strategy of Community Senior Activity Center Development

7.1 Introduction

China is rapidly aging. Affected by the family planning policy implemented in 1971, China shows characteristics of the nuclear family. This has gradually reduced the country's traditional home care capability, with the number of households composed of senior citizens with no children present increasing steadily. Therefore, the issue of senior care has increasingly become a social issue receiving widespread attention. Reliance on the social and community elderly care service system has gradually become the focus of elderly care issues. China has also issued policies and regulations to promote the development of "community elderly care." According to the regulations, each community must be equipped with a senior activity center consistently configured as required during construction. Thus, this paper explores seniors' use of community senior activity centers in the context of planning policy to identify possible problems and explores corresponding improvement measures.

Chinese research on community senior activity centers has focused mainly on architectural design, cultural construction, and physical layout, while there is a lack of micro-level research on user spatial needs and choices (Sun Rongwen 2008; Huixin 2013; Xing 2013). Uncertainties regarding seniors' activity center choices, which may conflict and influence each other during their actual selection and use, may cause analysis and research based on aggregate models to mischaracterize the complexity and uncertainty of seniors' behavior. Lack of clarity on the selection mechanism and demand for community senior activity centers may lead to such problems as imbalance between supply and demand in planning practice. China has issued a policy to ensure the construction of community senior activity centers, which states that "Communities should be equipped with community senior activity centers. The basic configuration of community-level senior activity centers includes activity rooms, reading rooms, health rooms, and an outdoor area of not less than 150 square meters. The construction area is 200–250 square meters per community

Y. Ma, Z. Shen, *Strategic Spatial Planning Support System for Sustainable Development*, Advances in Geographic Information Science,
https://doi.org/10.1007/978-3-031-07543-8_7

senior activity center. The community senior activity center can also be equipped with ball games room, chess and card room, gymnasium, leisure tables and chairs to enrich the life of the seniors." Thus, understanding how to analyze and predict the possible effects of existing planning policy is of great significance to future planning control and policy decisions. As an artificial intelligence method, agent-based modeling (ABM) has gradually become an effective technique to describe micro-level individual behavior mechanisms and to reflect the uncertainty and adaptability of individual behavior (Weiss 1999). Because it emphasizes the self-adaptive learning and selection process of individuals when interacting with the surrounding environment and other individuals, ABM has advantages in research on complex urban problems. In addition, because of its performance characteristics in temporal and spatial dynamic processes, this method has been applied to study urban individuals' independent behavioral choices, such as residents' responses to specific planning policies. Findings are then used to predict policy implementation effects and decision supports (Benenson 1998; He et al. 2013; Ma et al. 2013, 2016). However, in existing research, although interaction between agents is given great attention, strategies for using artificial intelligence methods to enable individuals to learn during the process of interaction remains unexplored.

Accordingly, this research applies artificial intelligence methods to construct an ABM for a simulation analysis model of the needs of community senior activity centers. In uncovering the interaction and mechanisms among the various elements in the use of these centers, the model serves to predict their needs and use patterns for a particular future period to assist in planning decisions. In the community facility for elderly (CFE) model, seniors can independently choose their daily use of an activity center and also adjust their behavioral choices spontaneously through learning based on the experience of using the center during interaction with other seniors. In summary, this work explores application of artificial intelligence in the planning and decision support of CFEs from the perspective of micro-level individual behavioral analysis.

7.2 Planning and Decision-Making Process for Community Senior Activity Centers Based on ABM

Using dynamic analysis and simulation methods to build an ABM involving the elderly, community senior activity centers and planners, this paper simulates and analyzes seniors' use of community senior activity centers in the context of existing norms. Based on the analysis results, this paper also conducts planning and control analysis of policy scenarios (Xing 2013; Hosseinali et al. 2013; Le et al. 2010), looking for feasible solutions to provide decision support for the planning and upgrading of urban community senior activity centers.

The planning and decision-making process of the CFE model is shown in Fig. 7.1. The process has two parts, simulation and visualization of the model and a

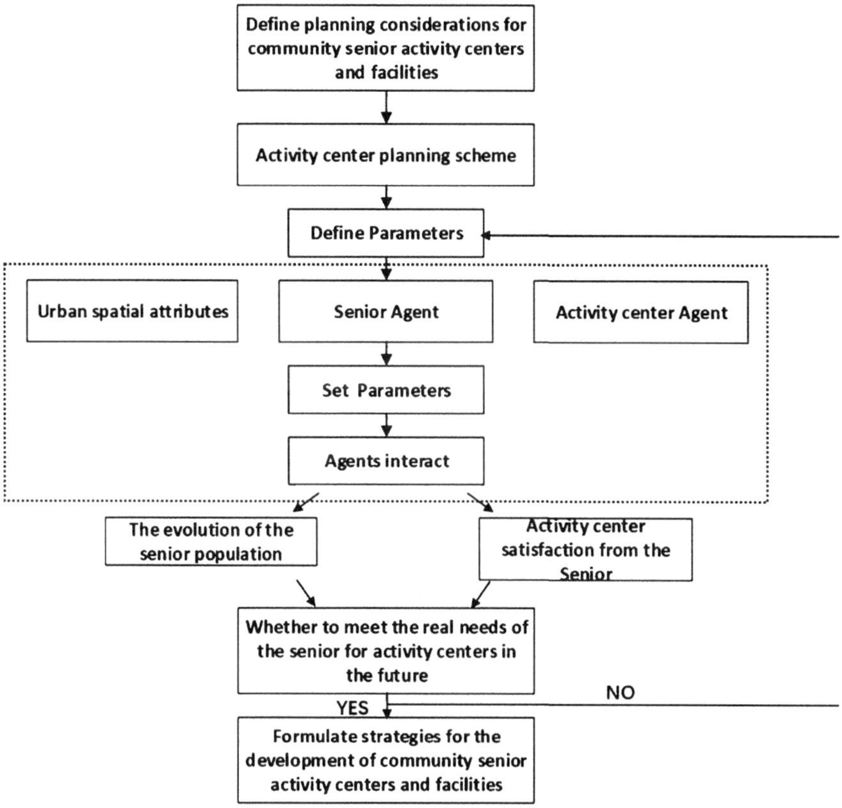

Fig. 7.1 Planning and decision-making process of community senior activity center based on ABM

participatory decision-making process. This paper first determines the strategic planning and layout of community senior activity centers through existing norms and sets this as the goal for the initial parameters of the simulation model. At the same time, the relevant data obtained from traditional survey statistics and the model parameters determined through research are used as the input conditions of the simulation. Through a background simulation operation, the prediction result is ultimately brought forth; this is the simulation output. The output results mainly include the population evolution of the study area and the level of seniors' satisfaction with their use of activity centers (Fig. 7.1). In Fig. 7.1, urban spatial attributes refer mainly to various natural, social and economic attributes of geographic space, such as land use status quo. These attributes comprise the foundation and carrier of the spatial characteristics of the senior agent and the activity center agent. When the senior agent chooses an activity center, this choice is affected by the geographic location of the center and its spatial attributes. Similarly, the location of the activity center is affected by such spatial attributes as whether the current land use is residential. Since China's activity center configuration has specific planning standards, it does not involve spatial location selection, so consideration of spatial attributes is relatively simple. The main indicators are the current status of land use in the study area and population size and spatial distribution.

7.3 CFE Model Design

7.3.1 Agent Design

An ABM that simulates and analyzes the space needs of a community senior activity center should be a smart model oriented to spatial planning and layout. The model should reflect the self-adaptive process of the research group's selection, evaluation and use of the community senior activity center, and the influence of surrounding groups and the environment. It can describe the dynamic process of population evolution as the simulation cycle progresses. The agent in the model is the active subject carrying out the demand for space in the community senior activity center, and the urban space is the spatial carrier for the survival and activity of these agents. In this model, the agent needs to have the ability to make independent decisions as well as the ability to interact with surrounding individuals and the environment. In view of the purpose of this research, we mainly discuss the spatial activities of three types of agents: the family agent, the CFE agent and the planner agent. Each agent has unique attributes and can exhibit individual behaviors independently. We generate adaptive behaviors and interactions among the agents via artificial intelligence algorithms.

7.3.1.1 Family Agent (Senior Agent)

In the CFE model, the family is regarded as an independent individual that makes independent decisions. Each family agent has attributes such as age structure, marital status, family members, income status and health status. Based on census data, this research predicts future changes in the number of seniors by determining the attributes of the family agent and simulating the life cycle changes of the family agent without considering mechanical growth of the population, as shown in Fig. 7.2.

7.3.1.2 CFE Agent

As the activity center's operator and manager, the CFE agent has attributes such as spatial environment level, facility level, management level and ability to accommodate seniors. The CFE agent is responsible for the activity center's construction, facilities and daily activities.

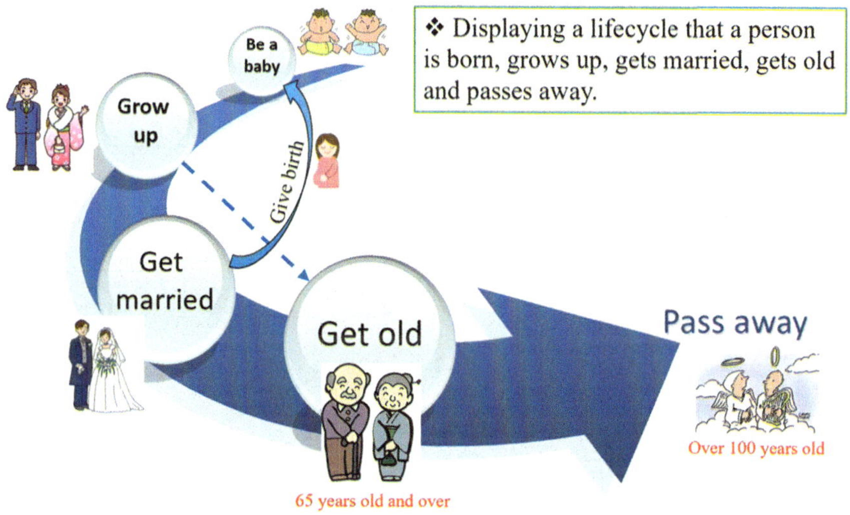

Fig. 7.2 Logic diagram of the life cycle

7.3.1.3 Planner Agent

In the CFE model, the planner agent is responsible for changing the level of supply of the activity center by adjusting the plan according to the number of seniors and their satisfaction with the built activity center during the simulation process. This shapes the future of the community senior activity center's plan and configuration to improve senior satisfaction with the activity center.

7.3.2 Agent Adaptive Behavior Selection Based on Individual Learning

7.3.2.1 Seniors' Choice of Community Senior Activity Centers

Seniors choose an activity center for leisure activities by clarifying whether they need to use the activity center and by evaluating whether the construction level of an activity center meets their own expectations. After the simulation program begins running, the senior agent selects an activity center according to the behavior rules shown in Fig. 7.3. The family agent will dynamically simulate the member structure and age structure over time. When the agent reaches retirement age, the agent's family members evaluate whether they need to use the activity center and they evaluate the effectiveness of the activity center. If the activity needs of seniors are met and the activity center can still accommodate additional people, the members will visit the activity center regularly in accordance with the distance factor.

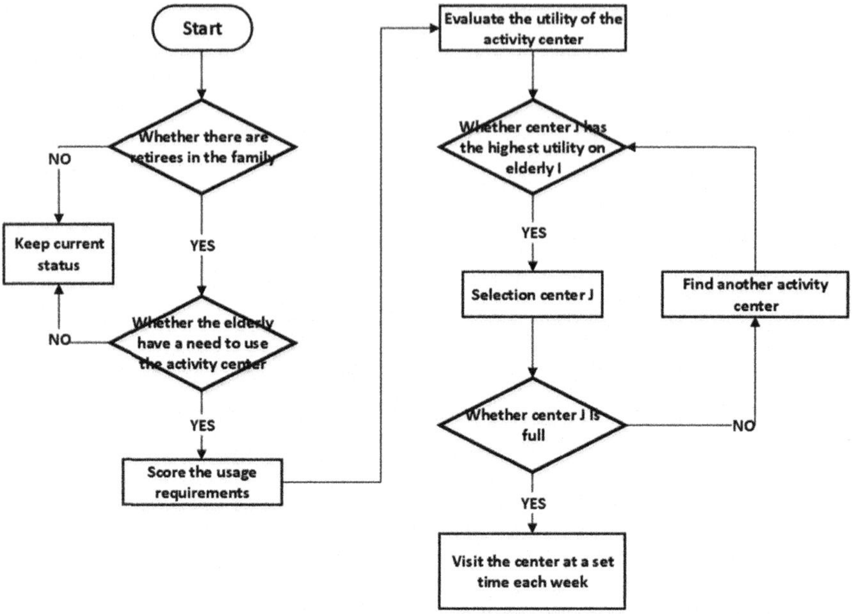

Fig. 7.3 Flow chart of selection of activity center by seniors

Due to differences in the preferences of seniors, the construction level of the activity center will or will not meet the activity needs of the seniors to varying degrees. Seniors' subjective preferences regarding spatial quality, facility level and service management level of the activity center and its actual planning and supply will jointly determine their satisfaction score for the activity center. At the same time, this research considers other negative effects, such as distance and depreciation level of facilities over time. If the overall construction level of the activity center meets the needs of seniors, they will give a satisfactory evaluation of the center that meets their own needs, calculated via the following equation:

$$S_{\mathrm{i}} = \sum_{\mathrm{j}=1}^{n} \left(\mathrm{b}_{\mathrm{ijs}} * X_{\mathrm{ijs}} \right) + \varepsilon_{\mathrm{i}} \ (n = 1,2,3,4)$$

where b_{ijs} is the preference coefficient of senior i to the jth influencing factor of activity center s (specifically including the three influencing factors of space quality, facility level and service management level). The value of x_{ijs} is senior i's score for the jth evaluation factor of activity center s. The sum of the products of the two values is the final score of the center. The values of ε_i refer to other related factors and are negative, including the distance from activity center i to a particular neighborhood, and the level of facility depreciation over time. This parameter can reflect the negative impact of the overall evaluation of the activity center.

7.3.2.2 Planning Management Decisions

Due to factors such as distance and differing preferences of seniors, the model will dynamically display the current visits to each center after each run, so that planners can intuit seniors' preferences and observe the satisfaction level of each center. At this time, the planner can intervene in planning decision-making (as shown in Fig. 7.4) and observe whether seniors' choices change. After the simulation is

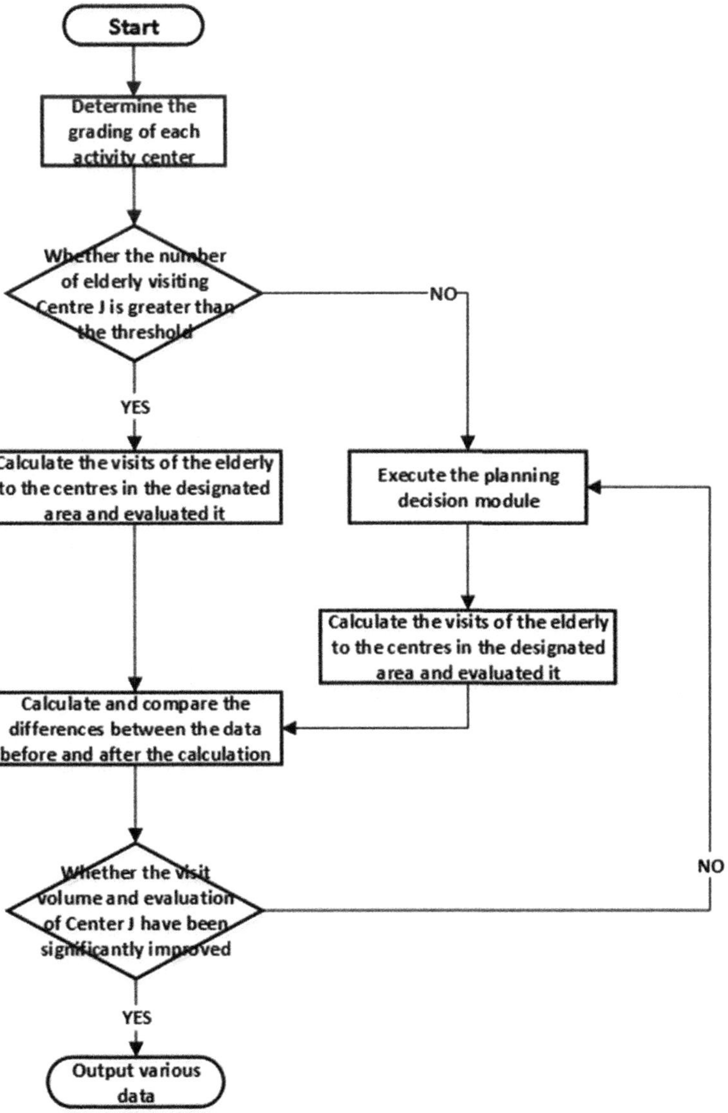

Fig. 7.4 Simulation process and flow chart of planning decision-making intervention

completed, real-time data is obtained, including changes in the preferences of seniors for the activity center based on different planning decisions and whether the decision significantly improves the use efficiency of the activity center. Data analysis can assist the planner in relevant decision making. After the policy is implemented (that is, the model parameters are changed and the specific parameters are changed according to different policies), the calculation of the model is repeated. The planner can compare the satisfaction status of the community senior activity center before and after implementation of the policy, so as to evaluate the effect of the policy implementation.

7.4 CFE Model Application—A Case Study in Downtown Fuzhou City, China

7.4.1 Model Development Platform, Data Pre-processing and Parameter Attribute Determination

7.4.1.1 Model Data Pre-processing

Section 7.3 describes how to build and simulate the model. This section presents a case study on the downtown area of Fuzhou to discuss verification and practical application of the CFE model. The CFE model was developed on the Netlogo platform created by Uri Wilensky in 1999. In terms of spatial data processing, first the raster data of the spatial attributes required for this study were obtained through rasterization of current land use data within the administrative divisions of Fuzhou. Data spatialization was then performed on location information for the residential neighborhoods obtained from the network to transform the data into points of interest in the community. Attributes include coordinate information, community name, community attributes and address information. The sixth census data for Fuzhou City were then spatialized to obtain the spatial distribution of population size and age structure. Finally, the above data were integrated and imported into the simulation platform, as shown in Fig. 7.5.

7.4.1.2 Determination of Model Parameters

The model parameters involved in this study include four main aspects: population changes, urban space attributes, family agent attributes and activity center agent attributes. The population change data include birth rate, marriage rate, and death rate. These data are used to deduce the life cycle model, which is used together with the small sample data of the previous survey to determine the proportion of seniors over 65 who need to use an activity center, and to further simulate and forecast changes in the number of seniors in the future.

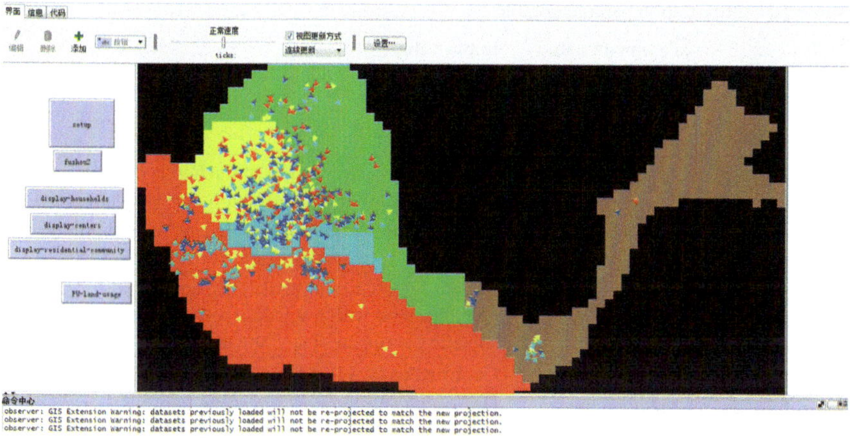

Fig. 7.5 Data visualization results of Netlogo platform. (Source: Self-drawn)

The initial parameters of community senior activity centers are based mainly on the land use planning of downtown Fuzhou and the *Urban Elderly Facility Planning Standards*. Combined with preliminary research, three initial parameters were determined. The only data that changes over time is population and community senior activity center data, so population spatial distribution in Fuzhou in 2010 was used as the initial population data for simulation. Because it is very difficult to implement a full-sample survey, a random sampling survey was adopted to collect statistical data on seniors in the downtown area of Fuzhou, including information on their activities and living conditions, and personal preferences. These data were used as a reference to set model parameters. In the model of the evolution of the population age structure, the life cycle parameters were based mainly on data from 2000 to 2010. The birth rate was set to 1.10%, the death rate to 0.90%, and the marriage rate to 86.0%.

Seniors have certain distance requirements when choosing an activity center. Therefore, the point-of-interest information of the residential neighborhoods obtained through the network was used to determine the coordinates and spatial location of each activity center. The number of households and community coordinates obtained through the network were used to determine the size and location of the senior population in each neighborhood. Because the micro-level behaviors and activities of seniors are biased, the model parameters should also consider individual differences of seniors. The factors that affect seniors' selection of an activity center are the attributes of the community and spatial location. Mathematical models are used to quantify seniors' demand for activity facilities. Parameter settings are shown in Table 7.1.

In terms of facility surveys, we took a typical Fuzhou community as an example. Considering the cultural characteristics of the popularity of chess and cards in Fuzhou, we conducted an in-depth investigation of the six necessary categories of activity facilities: chess and card room, reading room, activity room, gym,

Table 7.1 Initialization parameter table

	Parameter name	Initial value
	Birth rate	1.10%
	Death rate	0.90%
Global variable	Marriage rate	86.0%
Urban space attributes	Land use zoning	The administrative district of Fuzhou is divided into five areas to determine the distribution of residential land
	Family density	Set the number of households in each neighborhood according to the survey results for residential land
	Distribution of activity centers	According to current regulations, equip each neighborhood with a standard activity center and the specific coordinates (x, y) are determined
Family agent attributes	Location	Coordinates (x, y)
	Membership	Determine age composition of the family and average household population of each family based on survey results
	Senior preference	Preference for space, preference for facility level, preference for service quality, and requirement for distance
Activity center agent attributes	Capacity for seniors	Initialized standard activity center has a capacity of 30 people
	Activity center area	Initialized standard activity center has an area of 150 square meters
	Facility status	Facility level score of the initialized standard activity center is set to 3 points
	Service level	Service level score of the initialized standard activity center is set to 3 points
	Internal space environment	Internal space environment score of the initialized standard activity center is set to 3 points

multimedia room and health room. This investigation concluded that seniors of different age groups and education levels have different facility needs. Our analysis of the survey and the activity needs of the seniors led to the conclusion that the influencing factors that affect the use of activity spaces and facilities are spatial layout, facility conditions and service levels. These factors were used as parameter attributes to establish the ABM for the CFE.

Through a large number of preliminary investigations on the conditions in communities and for elderly families, we found that communities in the downtown area of Fuzhou can be divided into four types: educational communities, administrative communities, park communities and resettlement communities. Educational communities refer mainly to educational resources such as nearby schools. Administrative communities belong to the government's key construction communities. Resettlement communities refer to resettlement houses for those whose homes have been demolished. Leisure communities refer to parks or leisure squares. To a certain extent, community type indicates the differences in the attributes of the senior families living there, thus reflecting the different preferences of seniors for the activity center and different requirements for facility construction. This influences selection and evaluation of the facilities. Therefore, based on the preliminary research results, we use the decision tree shown in Fig. 7.6 to personalize seniors' preferences, so that the preferences of different seniors can not only meet the actual situation, but also be classified to the broadest degree.

To determine the demand value of the preference parameters of seniors, we investigate the basic situation of seniors in Fuzhou and their demand for space quality, facility level and service management level of community senior activity centers. A full-sample survey requires a lot of manpower. Therefore, the relationship between the type of senior and the demand value was obtained from a small sample of data. The demand value was then compared with the value of the proposed demand. A float value was used to represent the subtle differences between the preferences of different seniors, so that the simulation would be closer to the real situation. Seniors were asked to score three activity center construction projects and compare and rank the degree of importance they would attach to the three construction levels. The three scores summed to 100. The average scores for individual factors among the same types of seniors were obtained to facilitate calculation and conversion. As a result, the preference values of different types of seniors for various facilities were obtained (see Table 7.2).

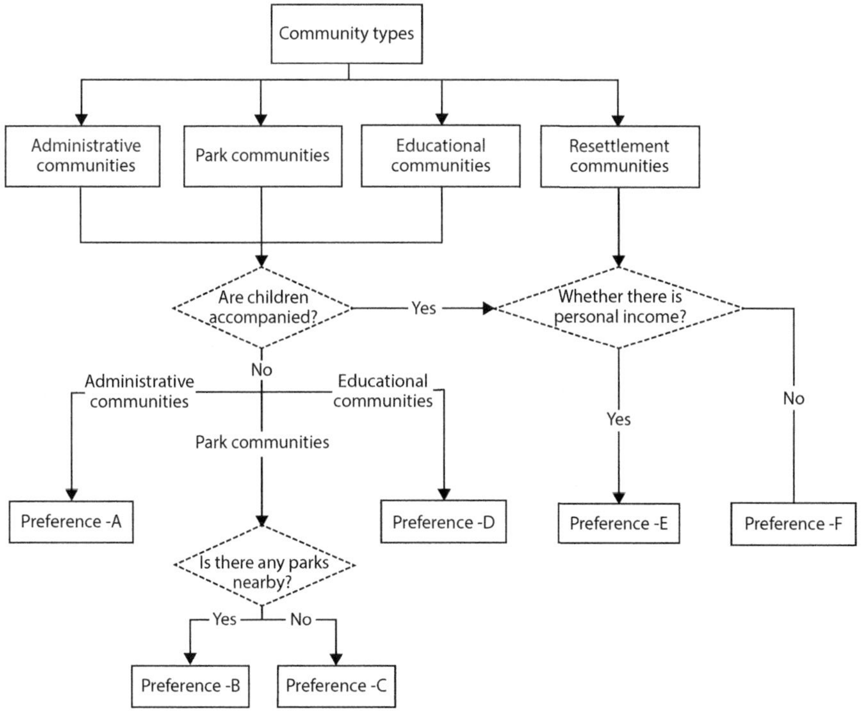

Fig. 7.6 Decision tree of preference judgment for seniors

Table 7.2 Seniors' six preference parameters

Preference name	Preference type	Parameter value
Preference-A	Seniors living alone in an administrative community	$demand_{1,2,3} = \{0.3,0.6,0.6\} + \{float1.8\}$
Preference-B	Social seniors living alone near the park	$demand_{1,2,3} = \{0.6,0.4,0.5\} + \{float1\}$
Preference-C	Social seniors living alone without a park nearby	$demand_{1,2,3} = \{0.5,0.4,0.7\} + \{float1.6\}$
Preference-D	Seniors living alone in an educational community	$demand_{1,2,3} = \{0.6,0.6,0.6\} + \{float1.6\}$
Preference-E	Seniors without preference who are accompanied by children and have independent income	$demand_{1,2,3} = \{0.7,0.6,0.5\} + \{float1.8\}$
Preference-F	Seniors without preference who are accompanied by children and have no independent income	$demand_{1,2,3} = \{0.3,0.2,0.4\} + \{float1\}$

In Table 7.2, $demand_{1,2,3}$ respectively represents the values of seniors' demand for space quality, facility level and service management level. The float value represents the subtle difference between the preferences of different seniors, making it more in line with a real situation.

7.4.2 Model Testing

7.4.2.1 Stability Test

Experiments on the model were repeated 10 times, each simulating 20 ticks (each tick representing 1 year, a complete model simulation cycle). For the 10 experiments, a comparison was made of the number of seniors and the number of visits to the activity center for each tick (see Fig. 7.7). Note from Fig. 7.8 that in the 10 repeated experiments, all initial parameters are kept unchanged, and the data for the senior population and the number of visits to the activity center obtained via simulation show little fluctuation, which conforms to the design principle of the simulation model.

7.4.2.2 Sensitivity Test

In the sensitivity test, first, the initial scores of the three evaluation factors were set at 3.5 points. The remodeling score was set to 0. From this, the relationship between center remodeling and satisfaction of the seniors was obtained. Seniors' satisfaction with the activity center fluctuated around 33%. Second, each parameter was set to 3.5 after reinitialization. That is, the initial attributes remained unchanged, and the

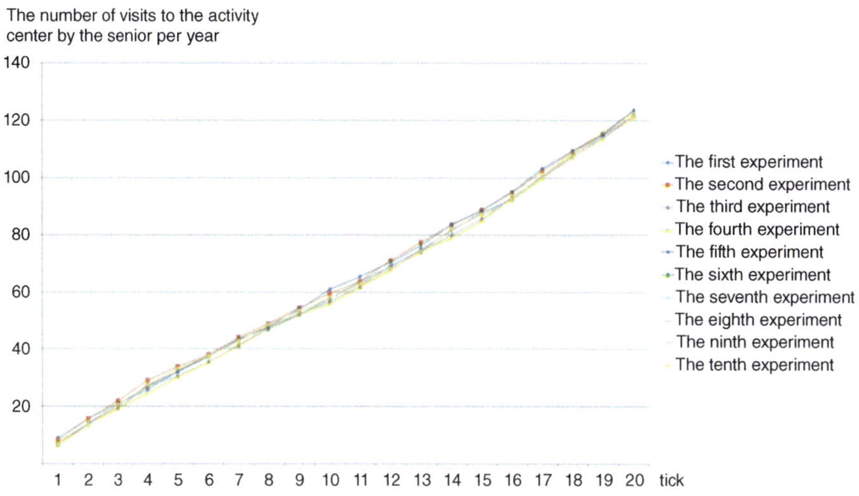

Fig. 7.7 Experimental results of model stability. (Source: Self-drawn)

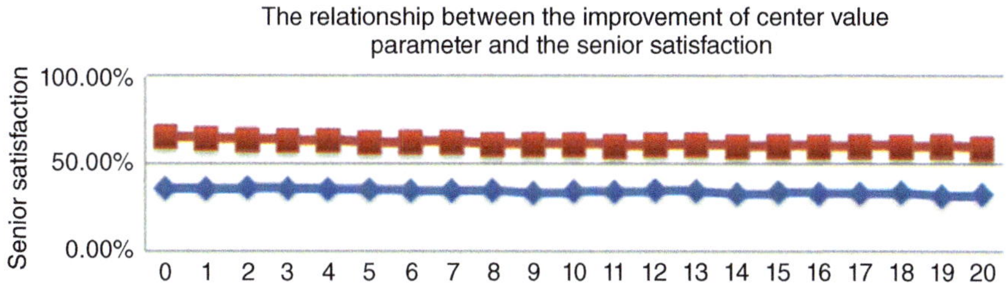

Fig. 7.8 Distribution of senior satisfaction after parameter modification. (The blue line represents senior satisfaction before the center is remodeled, and the red line represents senior satisfaction after the remodeling)

activity center with low evaluation scores was remodeled. The remodeling value was set to 6, thereby obtaining the relationship between the remodeling parameter and seniors' satisfaction.

The parameters of each experiment were adjusted once, the model was repeated 50 times, and the 50 experiments all showed stable simulation results, showing sensitive characteristics. A comparison was made of senior satisfaction from the two experiments, as shown in Fig. 7.8. Note that with basically no change to the population structure, senior satisfaction improved significantly after the remodeling value parameter of the poorly evaluated activity center was improved, fluctuating around 61%. This sensitivity test illustrates the validity of parameter values related to activity center evaluation. In the same way, the validity of other secondary parameters was verified in turn. Birth rate, death rate and marriage rate were found to be contributors to changes in the model simulation results; that is, each parameter is valid.

7.4.2.3 Verification with Real Data

In the process of verifying the model with real data, due to the huge amount of data for Fuzhou, to improve efficiency, we extracted several neighborhoods in Gulou District for data verification. A feasibility test of a small portion of the data for Gulou District served as a practical basis for the feasibility of the entire model. Historical population data for 2010 show that the total population of the community in 2010 was 30,000, and the corresponding senior population was 3268. Indicators for the community senior activity centers were set to 3.0. In accordance with our research logic, the satisfaction result after 6 years was obtained. Senior satisfaction with community senior activity centers ranges from 31% to 34%. This result from the model operation is consistent with the 32% found in the preliminary investigation; thus, the feasibility test of the model is reasonable (as shown as Table 7.3).

7.4.3 *Model Simulation Results*

7.4.3.1 Simulation and Prediction of Space Demand for Community Activity Centers in 2030

Using imported spatial data for Fuzhou City and the sixth census data for Fuzhou City, based on the current situation survey, this study predicts the supply and demand relationship between activity centers in Fuzhou and seniors till 2030. Figure 7.9

Table 7.3 Model simulation results of senior satisfaction

Number of years	Number of seniors	Number of years	Number of seniors	Satisfaction
0	3268	0	1058	32.37%
1	3453	1	1111	32.17%
2	3629	2	1147	31.61%
3	3780	3	1234	32.65%
4	3916	4	1225	31.28%
5	4070	5	1287	31.62%
6	4187	6	1441	34.42%

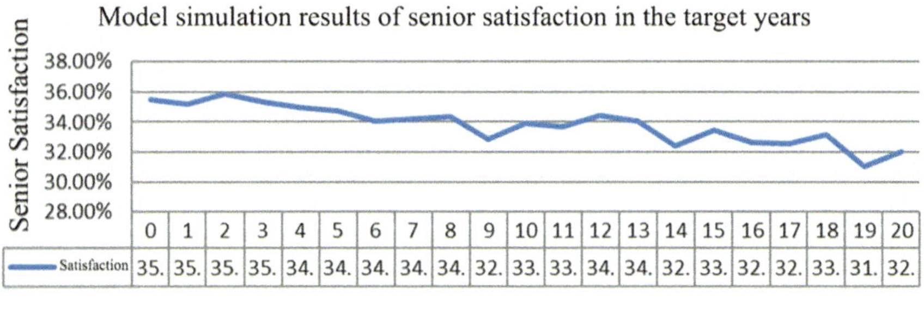

Fig. 7.9 Model simulation results of senior satisfaction in the target years

shows that in the next 10 years, senior satisfaction will show a fluctuating increase, and a downward trend will be more obvious. Compared with the staring year of 2010, satisfaction will have dropped by 3.45% by 2030. This means that under the current circumstances and construction level, resources are still not being used rationally.

7.4.3.2 Simulation for Policy Scenarios to Explore the Planning Strategy of Community Senior Activity Centers

In view of the problems in the simulation prediction results, we propose a variety of corresponding planning control programs and conduct scenario simulation and comparison for each policy program, leading to the best policy program. The model parameters are adjusted according to different policy scenarios. For a particular activity center, when there are fewer satisfied seniors than unsatisfied seniors, the center is displayed in red.

1. Policy Scenario 1: Government subsidizes activity centers with low satisfaction (a certain number of times) and expands those with high satisfaction without demolishing those with low satisfaction.

From the results (Figs. 7.10 and 7.11), note that this policy behavior will help improve satisfaction levels. In this scenario, it is predicted that the activity center satisfaction level could reach 77.47% by the target year.

2. Policy Scenario 2: Government subsidizes activity centers with low satisfaction (a certain number of times), puts a ban on those with which seniors are dissatisfied and uses funds to expand those with high satisfaction.

Note from the results (Figs. 7.12 and 7.13) that this policy behavior will help improve satisfaction levels. In this scenario, it is predicted that satisfaction with activity centers could reach 77.33% by the target year. However, Scenario 2 is not much different from Scenario 1, so demolishing centers with low satisfaction has little effect.

3. Policy Scenario 3: Overall improvement is made on individual items in senior activity centers (such as the internal space environment).

Taking improvement of the internal space environment of an activity center as an example, the internal space environment among the four attributes of the initial activity center was increased from 3 to 3.5, and the model run. Note from the results of the simulation prediction (Figs. 7.14 and 7.15) that overall improvement of the individual items can increase initial satisfaction with an activity center. Satisfaction trends upward, but the improvement is relatively slow. During the 20-year period, satisfaction level increases by only 13.81%.

According to simulation of the above three policy scenarios, overall improvement of the individual items can improve satisfaction with activity centers to a

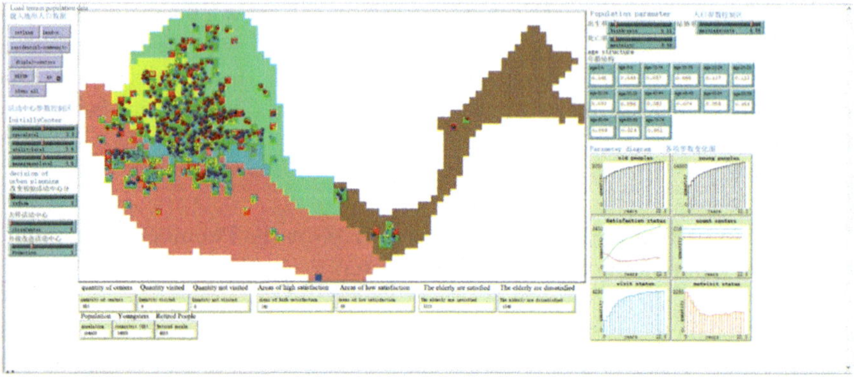

Fig. 7.10 Satisfaction results after intervention of Policy Scenario 1

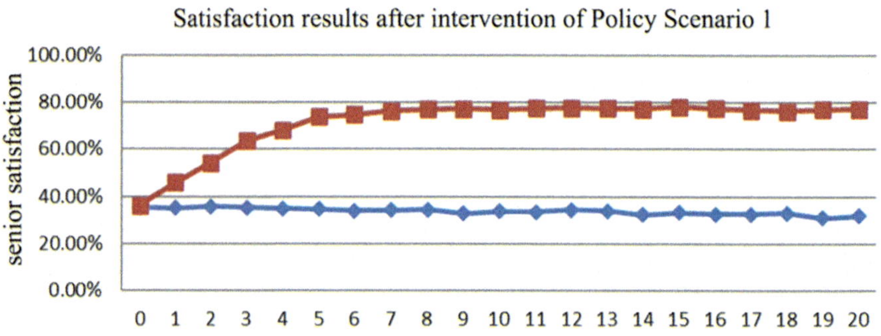

Fig. 7.11 Satisfaction results after intervention of Policy Scenario 1. (The red line represents satisfaction after intervention of Policy Scenario 1 and the blue line represents the original result)

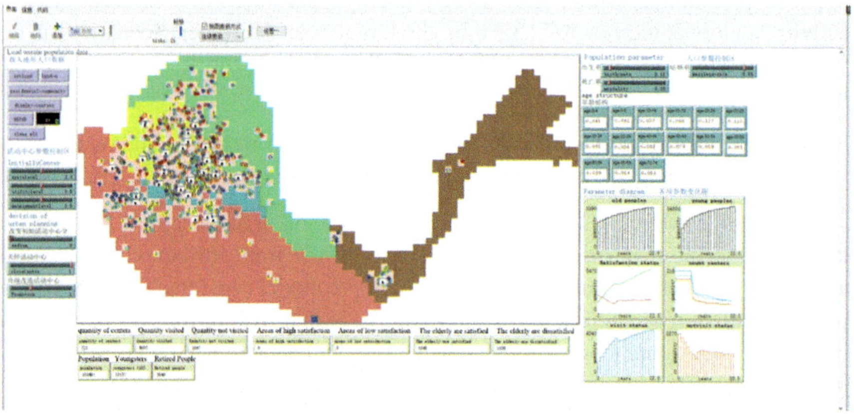

Fig. 7.12 Satisfaction results after intervention of Policy Scenario 2

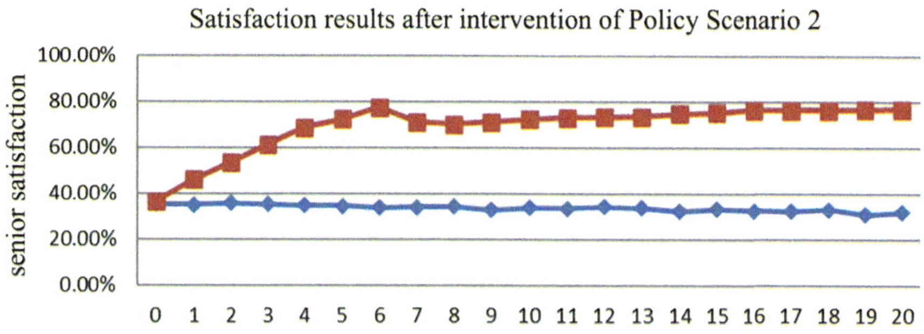

Fig. 7.13 Satisfaction results after intervention of Policy Scenario 2. (The red line represents satisfaction after intervention of Policy Scenario 2 and the blue line represents the original result)

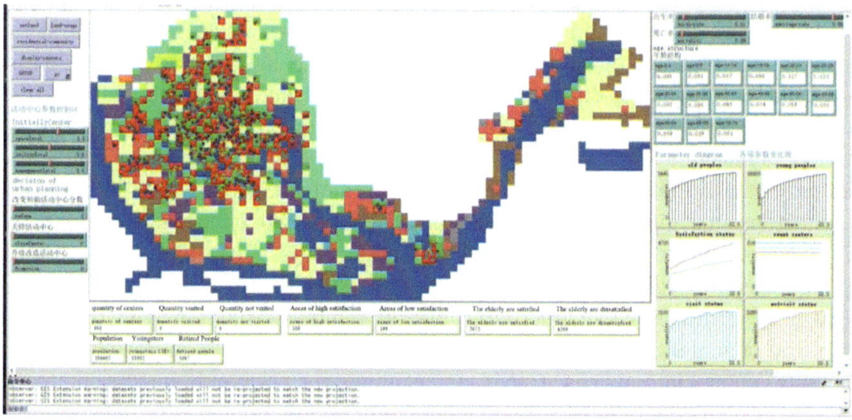

Fig. 7.14 Satisfaction results after intervention of Policy Scenario 3

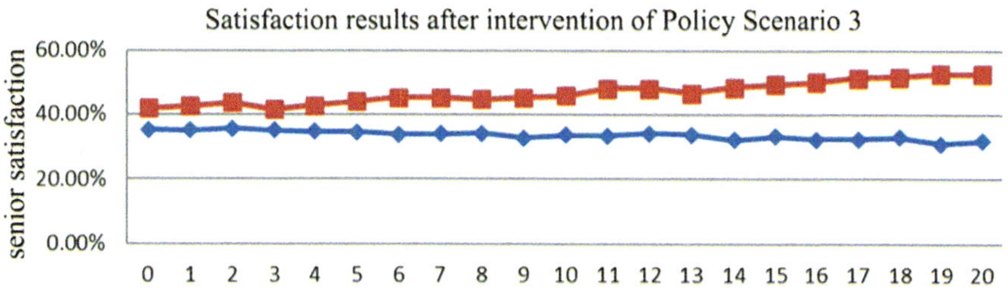

Fig. 7.15 Satisfaction results after intervention of Policy Scenario 3. (The red line represents satisfaction after intervention of Policy Scenario 3 and the blue line represents the original result)

certain extent, but overall satisfaction improves relatively slowly. In contrast, separate treatment of those issues with which seniors are dissatisfied can significantly improve satisfaction, so Policy Scenarios 1 and 2 may be more effective.

7.5 Conclusion and Discussion

In the context of the mismatch between supply and demand in community senior activity centers given China's rapidly aging population, this paper strengthens the research on micro-level individual characteristics, behaviors and mutual influences. To cope with complex adaptive systems such as a multi-agent system (MAS), we propose a method of data mining micro-level individual behavior patterns based on survey statistics and establish an artificial intelligence simulation analysis model (CFE) for the needs of community senior activity centers. Taking the downtown area of Fuzhou as an example, we evaluate the satisfaction of community senior activity centers over a 20-year period, put forward various scenarios for simulation of different plans, predict and evaluate the plans, and finally choose the best plan. This can help planners understand the real future needs of the senior population and better help them provide corresponding community activity centers and related facilities.

Actual application of the model in the downtown area of Fuzhou shows that the planning and layout of community senior activity centers under the current standards are not adequate, and urban resources are not fully utilized, which easily leads to low satisfaction among seniors. According to the results of the model simulation, as far as the communities in the downtown area of Fuzhou are concerned, as the simulation time approaches 2030, senior satisfaction with community senior activity centers will decrease significantly. Through analysis of different remodeling policy scenarios, it is found that expansion of existing centers with high satisfaction makes little difference in improving overall senior satisfaction with or without removal of centers with low satisfaction at the same time. Also, improvement of an activity center's internal space environment alone, such as improving the comfort of a center's internal spaces, does not significantly improve satisfaction. Therefore, in future planning practice, the standardized planning and configuration of community senior care centers is clearly unreasonable. To improve overall senior satisfaction, those centers currently with higher satisfaction should be expanded appropriately. However, whether removing those currently with low satisfaction will have minimal effect on improving overall satisfaction in the future. In addition, in contrast to the initial research assumption, simply improving the internal environment of a community senior activity center has no significant effect on improving senior satisfaction with the center. Therefore, overall remodeling of an individual item in an activity center should not be focused on the internal environment.

This study has limitations. For example, only the impact of improvement of the internal space environment of community activity centers on senior satisfaction is selected in Policy Scenario 3, and differences in service levels and facility conditions are not discussed individually. Thus, future research should further study the impact of service level, service content and facility status on senior satisfaction with activity centers. Building on the conclusions of this research, we should further explore activity centers with which seniors are more satisfied in terms of their service content, service level and facility characteristics. This would allow us to better understand planning and supply methods for community senior activity centers and improve policy regulation. Despite the shortcomings, construction and application of the CFE model shows that when faced with complex urban problems involving micro-level individual behavior differences, when predicting the future, statistical analysis that simply follows the patterns of the past cannot easily reflect the internal dynamic change processes of urban systems. Only when the interaction and mutual influence between the various complex elements involved in the problem are explored can a simulation more closely reflect real dynamic change. However, it should be clear that although model prediction results cannot be directly applied to the planning of a community senior activity center, a model nevertheless provides new analysis methods and more comprehensive information for the formulation of planning decisions, which helps to more intuitively reflect the implementation effect of such decisions, thereby contributing to the planning decision-making process.

References

Benenson I. (1998). Multi-agent simulation of residential dynamics in the city. Computer, Environment and Urban System, 22 (1): 25–42.

He J H, Liu Y L, Yu Y, etc. (2013). A counterfactual scenario simulation approach for assessing the impact of farmland preservation policies on urban sprawl and food security in a major grain-producing area of China, AppliedGeography, (37):127–138.

Hosseinali F, Alesheikh A A, Nourian F. (2013). Agent-based modeling of urban land-use development, case study: simulating future scenarios of Qazvin city[J]. Cities, (31):105–113.

Jin Xing. (2013). Research on Remodeling Design Elderly Community Care Space in Xiamen. Huaqiao University. 27–38

Le Q B, Park S J, Vlek Paul L G. (2010).Land use dynamic simulator (LUDAS): A multi-agent system model for simulating spatio-temporal dynamics of coupled human-landscape system 2. scenario-based application for impact assessment of land-use policies[J]. Ecological Informatics, (5): 203–221.

Sun Rongwen. (2008). Study on the Architecture Design of the Aged Entertainment in Urban Community——As an example for Beijing, Xi'an University of Architecture and Technology, 55–72.

Ma Y., Shen Z.J., Kawakami M. (2013). Agent-Based Simulation of Residential Promoting Policy Effects on Downtown Revitalization. The Journal of Artificial Societies and Social Simulation, 16(2):1–11.

Ma Y., Shen Z.J., Thanh D.N. (2016). Agent-Based Simulation to Inform Planning Strategies for Welfare Facilities for the Elderly: Day Care Center Development in a Japanese City. the Journal of Artificial Societies and Social Simulation, 19(4)5.

Weiss G. (1999). Multi-agent Systems: A Modern Approach to Distributed Artificial Intelligence. Cambridge, MA: The MIT Press, 105–108.

Wu Huixin. (2013). Study on the indoor and outdoor space design of city elderly activity center – Taking Beijing as an example. Beijing University of Civil Engineering and Architecture. 27–38

Chapter 8
Decision-Making Support System for Low-Carbon Urban Master Plan of New Town

8.1 Introduction

In previous chapters we introduced planning support systems for agents' behaviors in urban areas regarding different spatial strategic policies. In this chapter we will introduce how will urban master planning, as one of the strategic plan effects on urban environment. In Climate change has compelled governments around the world to implement low-carbon strategies. In China, with ongoing urbanization and economic development, cities, which are the main centers of greenhouse gas emissions, have aimed to undertake low-carbon city construction to meet the country's emission-reduction commitments. In this regard, urban and rural planning, which plays an important role in ecology, agricultural space protection, and urban layout, has become oriented toward attaining China's goal of hitting peak emissions and achieving carbon neutrality. As a result, the low-carbon planning and development of New Towns has received considerable attention. However, the main focus has been on energy-saving measures, which do not reflect emission-generating activities at the level of land type, making it impossible to take specific, quantifiable measures for different types of land.

To specify carbon emissions according to land type, it is necessary to establish a framework that correlates land use and carbon emissions – that is, establish the relationships between various emission-generating activities and specific types of urban and rural land. The Intergovernmental Panel on Climate Change (IPCC) refers to emissions from energy consumption caused by human activity on various types of land as indirect land-use carbon emissions. Emission sources can be divided into stationary and mobile sources, and land-use types corresponding to the related human activities are associated with those two types of sources. Against this background, this study aimed to build a correlation framework for land use and carbon emissions, involving an emission simulation and prediction model, emission

Y. Ma, Z. Shen, *Strategic Spatial Planning Support System for Sustainable Development*, Advances in Geographic Information Science,
https://doi.org/10.1007/978-3-031-07543-8_8

accounting method, and low-carbon assessment. On that basis, an intelligent decision-making system was developed for the low-carbon master planning of New Towns. This system can support decision-making for urban land-use planning of master plan, which is an important aspect of low-carbon planning in China.

Stationary emissions mainly come from direct (end-use) energy consumption by facilities such as residential, commercial, industrial, and government buildings. The carrying land for such emissions is relatively clear. Mobile emissions, meanwhile, come from traffic emissions, such as those generated by inter- and intracity passenger transport and freight transport. These activities are mainly carried by the areas used for roads, transportation facilities, airports, docks, ports, and so on. Their emissions are directly proportional to the scale of passenger and freight transportation, and the scale of freight transportation is directly related to land for urban logistics and storage. In China, the land use zones of urban master plan are classified into eight major categories. Accordingly, this study classified emission behaviors into eight land systems, including residential, administration and public service, commercial and business facilities, industrial, logistics and warehouse, street and transportation, municipal utilities, and carbon sinks. Among them, the emission behavior of residential land, commercial and public facilities, and industrial systems mainly comes from end-use energy consumption, such as production and building operations.

8.2 Carbon-Emission Simulation and Low-Carbon Assessment for Master Planning

8.2.1 Simulation Framework

This study's framework was based on a mechanistic planning scheme–environmental impact model. Here, it was necessary to first analyze the mechanism of the human process on the surface caused by the planning scenario and on that basis, construct a low-carbon simulation model of "New Towns" in China. The model aimed to achieve the purposes of quantitative analysis in three aspects: (1) the spatial layout and form, transportation system and land-use layout, and spatial distribution of residences and industries in the planning area under the action of spatial planning policy; (2) traffic volume and its carbon emissions caused by the spatial pattern of the planning area; and (3) total carbon emissions in the planning area based on the carbon emissions per unit of the planning land-use unit. Based on simulations of those three aspects, we evaluated possible future energy-saving and emission-reduction effects under the action of specific spatial planning policies and made adjustments and decisions about it.

This study's simulation model was mainly based on the implementation of master planning for New Towns in China. Such planning will bring about changes in the land use, spatial structure, spatial morphology, and transportation systems of the planning areas, in turn causing environmental impacts in the form of emissions. Environmental impact assessment was based on total carbon emissions as the measurement indicator. Through the carbon-emission accounting of different land-use types in the study area, carbon emissions were evaluated under the premise of certain planning policy inputs. After comparing the carbon-emission estimation result output after the input of various spatial planning policy options, the spatial planning policy parameters were identified under the optimal energy-saving and emission-reduction mode, thus providing decision-making support for spatial planning policy. Figure 8.1 shows the framework of the research model.

The input layer in Fig. 8.1 mainly refers to the planning policy that leads to the abovementioned land-use changes, which are mainly built around the constraints that need to be considered for land-use allocation in future planning. The impact layer concerns the development potential of various land types in the planning area for different functional layouts in the abovementioned allocation process; it reflects land-use and spatial-pattern changes in the planning area caused by the planning policy. Finally, based on changes in the functional layout and spatial pattern of the land, the total amount of carbon emissions was calculated, and the low-carbon level was judged, and these were fed back into decision-making for spatial planning policy.

8.2.2 *Model Simulation Design*

8.2.2.1 Prediction and Identification of New Towns' Spatial Structure Based on Land-Use Layout

Based on the land-use layouts determined by master planning, the model can predict and identify possible future spatial structures and various functional areas. This is used as a basis to support the subsequent forecasting of intracity traffic behavior in the planning area. The city center was divided into main center and subcenter, which

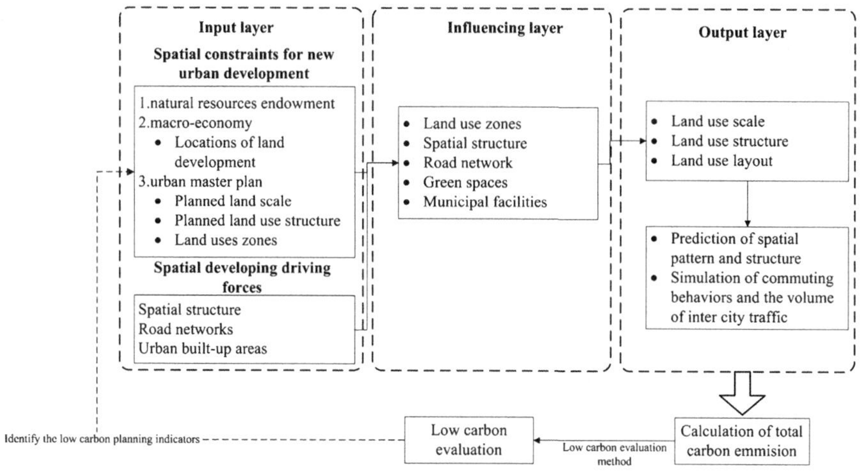

Fig. 8.1 New Town carbon-emission simulation framework

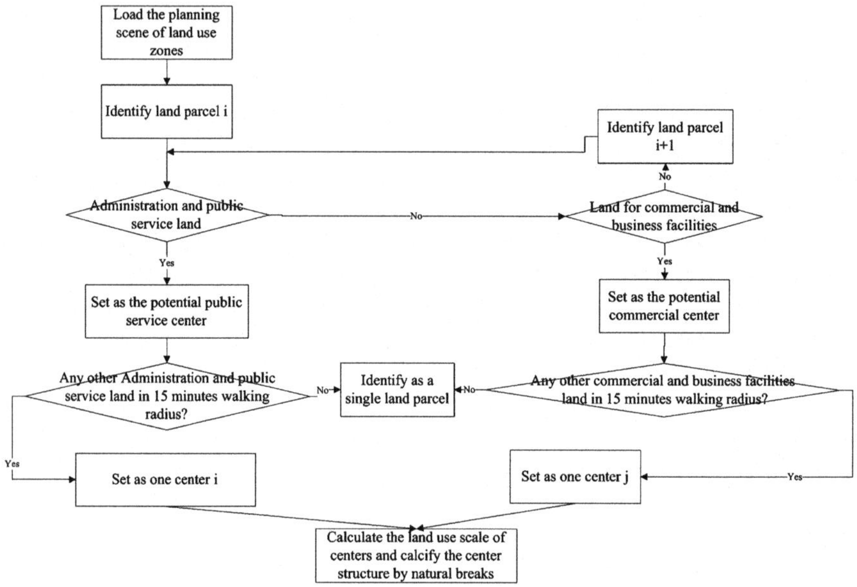

Fig. 8.2 New Town spatial structure prediction identification process

were specifically divided into two categories: public service and commercial center. Figure 8.2 depicts the logic of prediction and identification.

8.2.2.2 Intracity Traffic Origin–Destination (OD) Prediction Based on Traffic Behavior

The same land-use structure under different land-use layouts will produce different urban spatial structures and morphologies, which in turn will affect residents' transportation patterns and the volume of motor vehicles and have a greater impact on emissions from intracity transportation. In China, the land use zones of urban master plan are classified into eight major categories, which are residential, administration and public service, commercial and business facilities, industrial, logistics and warehouse, street and transportation, municipal utilities, green space. Therefore, based on simulating emissions that could be generated by the spatial layout of the eight major land-use categories in the New Towns' master plan, we further distinguished transportation emissions into two parts: carbon emissions generated by external transportation and those generated by intracity transportation. Emissions from urban transportation were calculated based on simulation and prediction of CO_2 volume generated by travel behaviors of planning population in the New Town. Commuting was divided into two types: commuting to work and commuting to school. Traffic volume prediction was based on identifying the traffic OD relationship. In the calculation process for commuting volume, we assumed the origin point

(point O) of a resident's traffic activity was the person's residence, and the travel destination (point D) prioritized the community living circle with place of residence as the origin. Therefore, when simulating travel destination, each resident was given priority to choose a travel destination in the living circle near his or her residence location (the planned residential zones). When the living circle could not be satisfied, the search was carried out to a farther radius. Figure 8.3 shows the specific simulation logic.

8.2.3 Building a Land Use–Carbon Emission Correlation Framework

Widely used carbon-emission accounting methods include the emission-factor, mass-balance, and actual-measurement methods, issued by the IPCC and other organizations. The emission-factor method is the mainstream accounting method.

We sorted out the relationships between existing domestic and international urban greenhouse gas inventories and urban master plans, integrating the scope, boundaries, and elements of emission-inventory accounting from the perspective of spatial planning. On that basis, we built a land use–carbon emission correlation framework based on classifications of urban and rural land, departmental classifications of urban greenhouse gas emissions inventories, and spatially planned emission-reduction paths. The framework establishes emission calculation and analysis methods for each of the following four modules: residence, commerce and public facilities, industry, and transportation. Thus, the emission intensity and total emissions of each module were calculated.

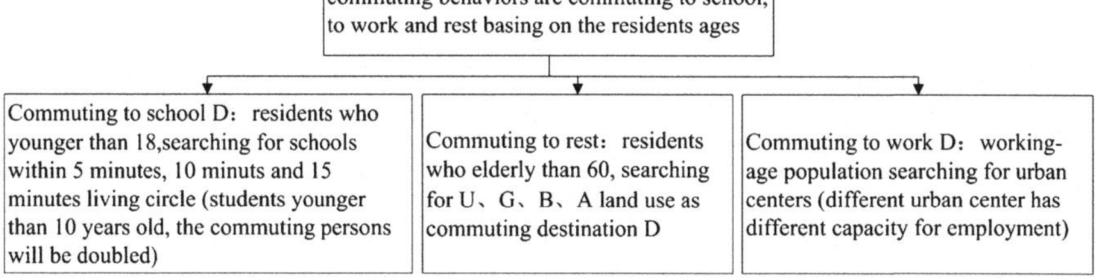

Fig. 8.3 Flow of OD simulation of residential traffic based on the land-use layout of the New Town

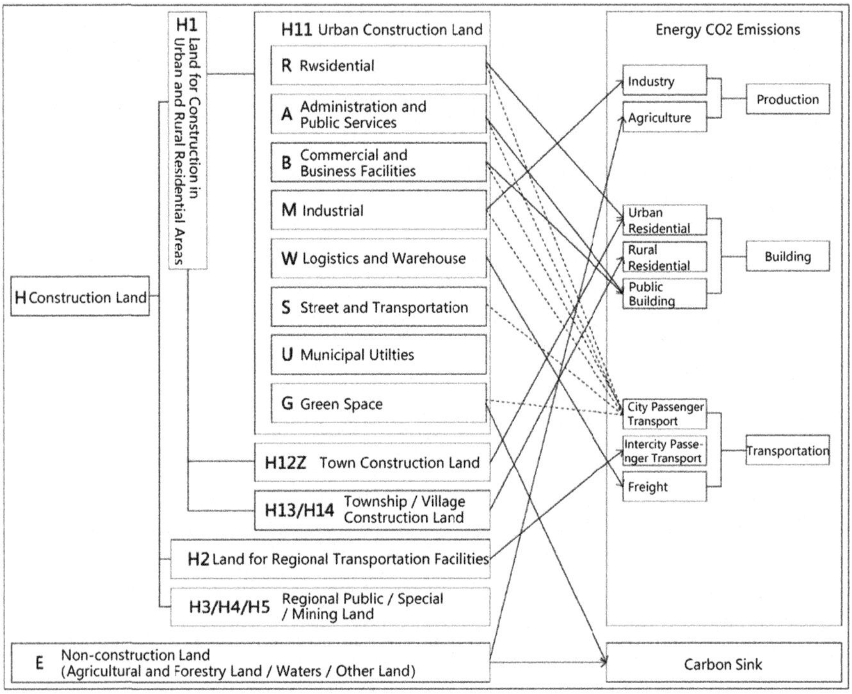

Fig. 8.4 Land use–carbon emission correlation framework

The land use–carbon emission correlation framework was established to correlate various activities that generate carbon emissions with urban and rural land-use types. Based on the literature, carbon emission sources were divided into stationary sources and mobile sources; land-use types corresponding to related socioeconomic activities were associated with those two types of sources. Figure 8.4 shows the details.

Stationary emission sources include direct (end-use) energy consumption in buildings and facilities, including residential, commercial, and industrial buildings and facilities, where emissions mainly come from end-use energy consumption, such as production and building operations. Mobile emission sources include energy consumption from transportation within or across city boundaries, which can be divided into intercity passenger transport, freight transport, and intracity passenger transport. Emissions from those sources are proportional to the scale of passenger and freight transport, and the scale of freight transport is directly related to urban logistics and warehouse. All types of carbon-emission behaviors were classified and correlated to the eight major land use types we introduced in Sect. 8.1.

8.2.4 *Carbon-Emission Accounting Method*

8.2.4.1 Calculating Carbon Emissions in New Towns

According to the carbon-emission intensity of different land types, the total carbon emissions of various land types can be summed to obtain the total carbon emissions, as follows:

$$C_{total} = \Sigma\left(P_i S_i\right) \tag{8.1}$$

where C_{total} is the total CO_2 emissions; P_i is the carbon-emission intensity of the ith type of land use zones; i is residential land, land for commercial and public facilities, and industrial land; and S_i is the area of ith type of land use zone. Land area can be obtained from vectorized planning drawings, and the carbon-emission intensity P_i of the land needs to be calculated based on the city's CO_2 emissions and land area in the benchmark year.

8.2.4.2 Accounting Methods for Carbon Emissions from Different Systems

The carbon-emission accounting methods for each type of land system are detailed below.

(1) Residential land: CO_2 emissions mainly come from various types of energy consumed by buildings and residential activity:

$$CO_{2RM} = \sum_{i=1}^{n}\left(k_i \times E_i \times d\right)_i \tag{8.2}$$

$$\mathrm{P}_{RM} = CO_{2RM} / S_{Nm} \tag{8.3}$$

$$CO_{2RM} = P_{RM} \times S_{Pm} \tag{8.4}$$

where CO_{2Rm} is the total CO_2 emissions from residential land in category m in the benchmark year; k_i is the coefficient of energy conversion to standard coal; i represents natural gas, liquefied petroleum gas, coal gas, electricity, and coal; E_i is the amount of energy consumed in category i; d is the emissions coefficient (fixed value); S_{Nm} is the residential land area of category m in the benchmark year; S_{Pm} is the residential land area of category m in the master plan; P_{Rm} is the carbon-emission intensity of residential land in category m; m represents residential land in categories 1, 2, and 3; and CO_{2RPm} is the CO_2 emissions of category m residential land in the master plan. We used domestic energy consumption from the statistical yearbook as the energy consumption of each endpoint of the residential system. Commercial and business facilities, industrial, logistics and warehouse, street and transportation, municipal utilities, green space

(2) Land for administration and public service: The calculation method for CO_2 emissions was the same as for residential land.

(3) Industrial land: CO_2 emissions are mainly generated by the energy consumption of various industries:

$$CO_{2ImN} = \sum_{i=1}^{n} \left(k_i \times E_i \times d\right)_j \tag{8.5}$$

$$\mathrm{P}_{IM} = CO_{2ImN} / S_{ImN} \tag{8.6}$$

$$CO_{2\mathrm{ImP}} = P_{Im} \times S_{ImP} \tag{8.7}$$

where CO_{2ImN} is the total CO_2 emissions from industrial land in category m; S_{ImN} is the industrial land area in category m in the benchmark year; S_{ImP} is the industrial land area in category m in the master plan; P_{Im} is the carbon-emission intensity of industrial land in category m; m represents industrial land in categories 1, 2, and 3; j represents industrial subsectors such as iron and steel; k_i is the coefficient of energy conversion to standard coal; i represents natural gas, liquefied petroleum gas, coal gas, electricity, and coal; E_i is the amount of energy consumption of type i; and CO_{2ImP} is the CO_2 emissions from industrial land in category m in the master plan.

(4) Street and transportation land: CO_2 emissions mainly come from intercity transportation and intracity passenger transportation. The former includes passenger and freight transport. Therefore, the emissions of transportation land were divided into two parts: intercity transportation emissions and intracity transportation emissions.

   (1) Intercity transportation emissions

   Previous studies investigated the carbon-emission coefficients per unit mile of four modes of intercity transportation in China: railway, highway, water transportation, and aviation. Emissions from domestic intercity passenger and freight transportation were found to account for 40% of total transportation emissions.

$$CO_{2\mathrm{T}} = CO_{2\mathrm{T1}} + CO_{2T2} \tag{8.8}$$

$$CO_{2\mathrm{T1}} = \Sigma\left(g_i \times h_i + g_{2i}\right) \times k_i \tag{8.9}$$

$$CO_{2T2} = \delta CO_{2\mathrm{T1}} \tag{8.10}$$

where CO_{2T} represents the total CO_2 emissions of the transportation system; CO_{2T1} represents the total CO_2 emissions of intercity transportation; CO_{2T2} represents the total CO_2 emissions of intracity passenger transport; g_{1i} (10,000 ton km) represents the passenger turnover of type i transportation; g_{2i} (10,000 km) represents the freight turnover of type i transportation; k_i represents the passenger-to-freight conversion coefficient; i represents railway, aviation, highway, and water transportation; and δ represents the carbon-emission ratio coefficient of intercity and intracity transportation.

(2) Intracity transportation emissions

Based on the intracity traffic OD prediction logic introduced in Sect. 8.2.2, we estimated traffic emissions using the total-structure method based on simulating the OD relationships in New Towns. The calculation method is detailed below.

First, calculate the total annual travel volume by population size, per capita travel times, and duration:

$$D = 365 \times P \times N \tag{8.11}$$

where D is the total annual trip volume, P (person) is the population number, and N (trips/day) is the number of trips per capita per day.

Second, based on the traffic volume calculation, emissions from intracity traffic are calculated as follows:

$$Emission = \sum_j \sum_i \left(\frac{D \times M_{i,j} \times d_{i,j}}{O_{i,j}} \times F_{i,j} \right) \times EF_j \tag{8.12}$$

where i represents the type of intracity transportation vehicle (e.g., taxis, private cars, buses); j represents different fuels (e.g., gasoline, diesel, natural gas); D is the total annual travel volume; $M_{i,j}$ is the travel sharing rate using fuel j of the i transportation type; $d_{i,j}$ (km/time) is the average travel distance using fuel j of the i transportation type; $O_{i,j}$ (person) is the average load using fuel j of the i transportation type; $F_{i,j}$ is the average energy consumption using fuel j of the i transportation type; EF_j (kgCO$_2$/kg) is the CO_2 emissions coefficient of fuel j; and Emission (kg) is the annual CO_2 emissions of intracity traffic in kilograms (kg).

8.2.5 Low-Carbon Assessment of Planning Schemes

Researchers have two general categories of evaluation methods for low-carbon cities, which are the main indicator method and the composite indicator method. Methods for low-carbon city indicator systems include subjective, objective, and combined methods. Commonly used subjective analysis methods include expert surveys and hierarchical analysis. Commonly used objective analysis methods include principal component analysis, DPSIR, entropy value, and TOPSIS. Combined analysis methods combine the abovementioned subjective and objective methods as the basis for empirical research and evaluation, taking into account of decision-makers' preferences. Some researchers, however, believe that such approaches combine methods that are not very relevant, making them meaningless for low-carbon city assessment. Among them, subjective analysis methods were first applied in research and have developed more maturely, while objective analysis methods have been more widely used in recent research.

For this study, we built an index system (Table 8.1) and on that basis used the expert scoring method to evaluate low-carbon master planning in New Towns. The weights of the five measurement indicators in Table 8.1 were generated by the weights of 16 secondary indicators contained. The formula for the low-carbon index of New Towns' master planning is

$$Z = \sum_{i=1}^{N} w_i y_i, (N = 1,2,\ldots,16) \tag{8.13}$$

where Z is the low-carbon evaluation index of the New Towns' master plan, Y is the score of index i, W_i is the weight of index Y_i, and N is the number of indexes.

The weight assignment used analytic hierarchy process (AHP) to mark the evaluation of the low-carbon space structure of the New Towns' master plan as target level A. Then, the target level was divided into five criteria layers – that is, S, L, T, I, and E – and several indicator layers below the criterion layer. For example, S was

Table 8.1 Low-carbon degree measurement indicators for New Towns' master plan

Standard layers	Indicator layers	Unit
Compact development degree (S)	Central equilibrium degree (S1)	–
	Gini index of population distribution (Gini) (S2)	–
	Space compactness (S3)	–
Mixed land use degree (L)	Land use balance (L1)	–
	Function mix (L2)	%
	TOD land layout (L3)	–
Public transportation orientation degree (T)	Bus station coverage rat (T1)	–
	Integrated road network density (T2)	–
	Proportion of slow road space (T3)	–
Facility constraint degree (I)	Green energy (I1)	%
	Integrated corridor (I2)	%
	Non-conventional water utilization rate (I3)	%
	Waste utilization rate (I4))	%
Ecological carbon sequestration degree (E)	Regional green space index (E1)	m^2/person
	Ecological control (E2)	–
	Greening index (E3)	m^2/person

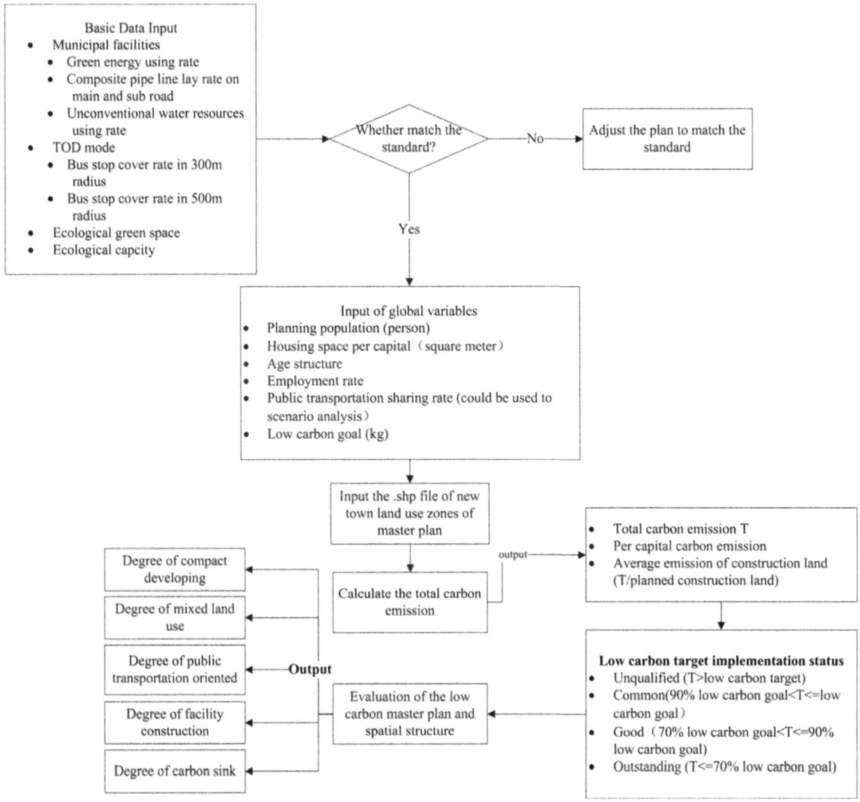

Fig. 8.5 The logic of decision-making support system for low-carbon master plan in New Town

divided into three indicator layers – S1, S2, and S3 – and other standard layers are similar, as shown in Table 8.1. Finally, we built the judgment matrix; used expert surveys to investigate the importance of the indicators, factors, and subfactors; and obtained a comparison judgment matrix by comparing the importance of the unified level evaluation items two by two. Primary and secondary indicator weights were obtained after calculating the consistency indicators by normalizing and reassigning the weights.

8.3 System Development and Testing

8.3.1 System Framework

Users can enter ArcGIS_ArcMap graphical data in the login interface, which contains the basic data and global variables for scenario analysis. The basic data are directly used for the low-carbon assessment of the New Towns' master plan. If the low-carbon measurement requirements are met, the simulation accounting of the total emissions of the New Towns' master plan and the low-carbon level assessment are carried out by setting global variables. Figure 8.5 shows the detailed process and logic.

8.3.2 Development Results

Based on the logic shown in Fig. 8.5, we integrated the development of a New Towns' low-carbon decision-making system. Figure 8.6 shows the specific operation interface, including its four major functional sections: the login interface, basic

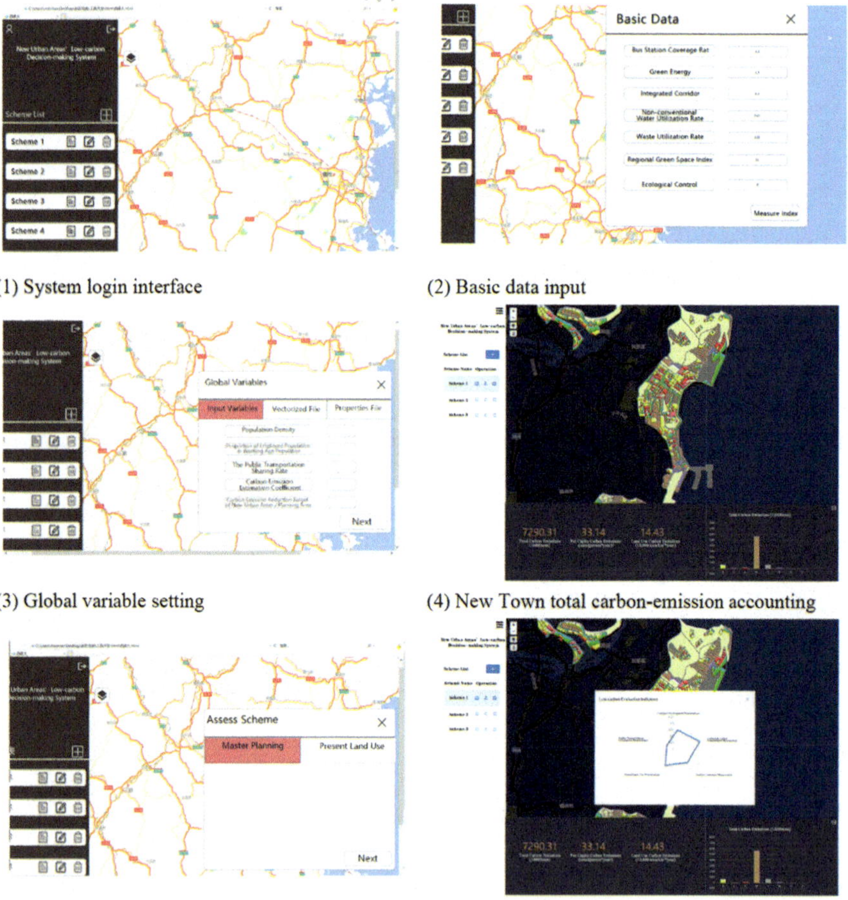

Fig. 8.6 The interface and operation modules of the decision-making support system. (1) System login interface. (2) Basic data input. (3) Global variable setting. (4) New Town total carbon-emission accounting. (5) New Town master plan low-carbon assessment. (6) New Town master plan low-carbon assessment results

data input, global variable input for scenario analysis, and carbon-emission accounting and low-carbon assessment. To support master-plan optimization, we focused on possible emission scenarios based on various planned built-up lands and their spatial layouts in New Towns. The specific built-up land classifications were consistent with the primary classifications in China's *Urban Land Classification and Planning Built-up land Standard, GB50137-2011* (hereafter, the Standard). The proportion of each type of built-up land was in line with the requirements of the Standard, as shown in Table 8.2.

8.3.3 System Testing

System testing was carried out as follows. First, for land layout and the land-use structure involved in the New Towns' master plan, we tested whether the system could carry out the corresponding logical simulation estimation for different layout schemes, land-use structures, and the corresponding total emissions. Second, we tested the global variables of the system, such as changes in the public transportation sharing ratio in relation to total carbon emissions and the low-carbon level of the New Town.

(1) Sensitivity and validity testing of the system for the land structure and layout of the New Towns' master plan

Here, we take the Fuzhou New Town as an example. By changing the land-use structure and adjusting the spatial layout of the planning, the carbon-emission prediction of different planning schemes was carried out to test the sensitivity and feasibility of the system's adjustment of schemes and the prediction of carbon-emission results.

Table 8.3 shows the structure of planned construction land use of Fuzhou New Town in scheme 1, and Fig. 8.7 shows the carbon-emission prediction result for scheme 1. The total carbon emissions are 72,903,100 tons/year, the per capita

Table 8.2 Built-up land classification and proportion control standards in the evaluation system

Land use classification	Proportion of urban built-up land (%)
Residential, R	25–40
Commercial and business facilities, B	
Administration and public services, A	5–8
Industrial, M	15–30
Green space, G	10–15
Street and transportation, S	10–25
Logistics and warehouse, W	
Municipal utilties, U	

Table 8.3 The structure of planned construction land use in scheme 1

Land use code	Land use type	Land area (km^2)	Proportion of built-up land (%)
R	Residential	105.02	20.78
A	Administration and public services	19.67	3.89
B	Commercial and business facilities	37.48	7.42
M	Industrial	98.49	19.49
S	Street and transportation	138.12	27.34
W	Logistics and warehouse	23.94	4.74
G	Green space	79.69	15.77
U	Municipal Utilties	2.85	0.56
H11	Urban built-up land	505.26	100.00

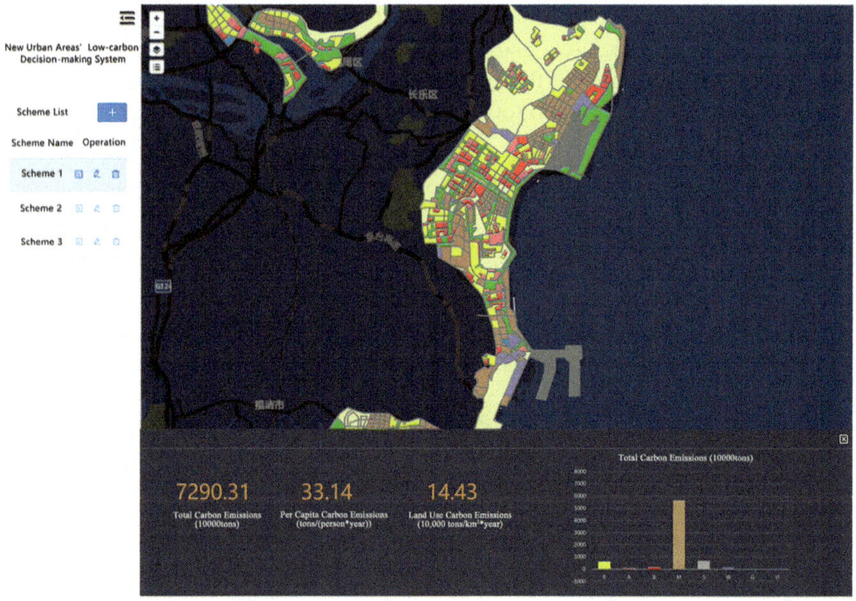

Fig. 8.7 Carbon-emission prediction results for scheme 1

carbon emissions are 33.14 tons/(person*year), and the average land carbon emissions are 144,300 tons/(km^2*year). The low carbonization of the New Towns' master-plan scheme was evaluated in five aspects: compact development, mixed land use, ecological carbon sequestration, facility constraint, and public transportation orientation. The evaluation showed that public transportation orientation was the main shortcoming of the Fuzhou New Town master plan. Aside from that, the results for the other four aspects were relatively high, as shown in Fig. 8.8.

Fig. 8.8 Low-carbon degree results for scheme 1

Table 8.4 Urban built-up land balance sheet in scheme 2

Land use code	Land use type	Land area (km^2)	Proportion of built-up land (%)
R	Residential	105.02	20.78
A	Administration and public services	19.67	3.89
B	Commercial and business facilities	37.48	7.42
M	Industrial	123.75	24.49
S	Street and transportation	112.86	22.34
W	Logistics and warehouse	23.94	4.74
G	Green space	79.69	15.77
U	Municipal Utilties	2.85	0.56
H11	Urban built-up land	505.26	100.00

Based on scheme 1, the first adjustment to land structure and layout was carried out to generate scheme 2. Scheme 2 mainly adjusts the land structure of Fuzhou New Town as follows: increase the proportion of industrial land by 5%, and reduce the proportion of road and transportation facilities by 5%. Table 8.4 shows the adjusted built-up land balance table. The plan after adjustment was reentered into the system for carbon-emission prediction and low-carbon degree evaluation. Figure 8.9 shows the carbon-emission prediction results for scheme 2. The total carbon emissions are 86.143 million tons/year, the per capita carbon emissions are

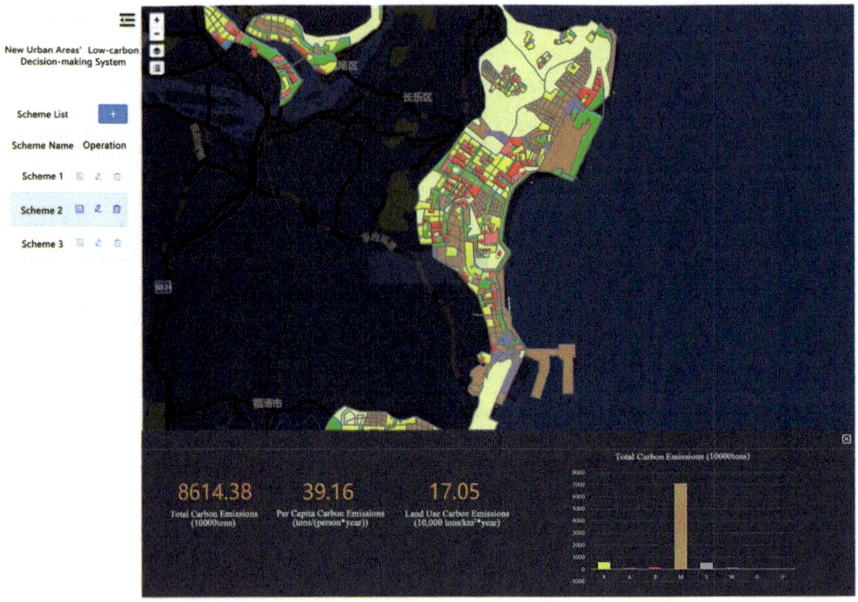

Fig. 8.9 Carbon-emission prediction results for scheme 2

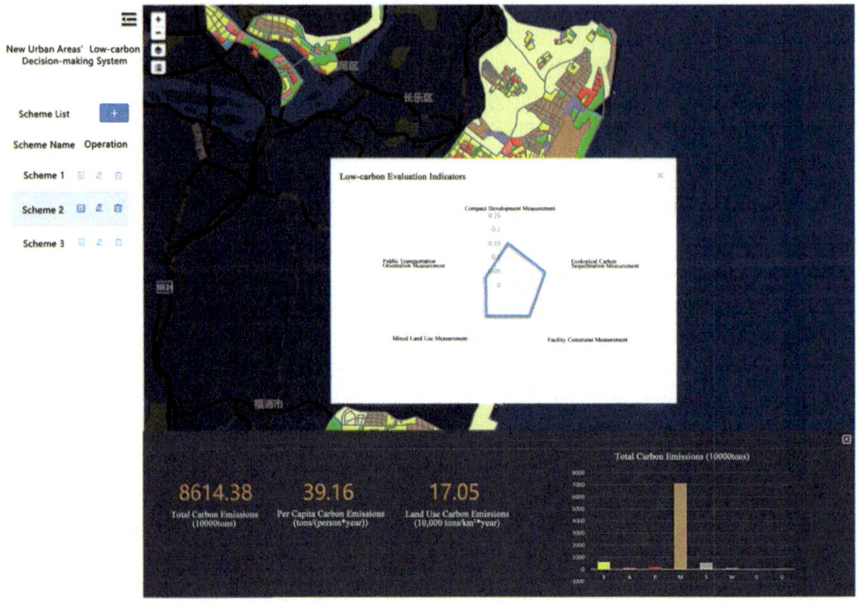

Fig. 8.10 Low-carbon degree results for scheme 2

39.16 tons/(person*year), and the land average carbon emissions are 170,500 tons/(km^2*year). Compared with scheme 1, the carbon emissions of industrial land in scheme 2 are significantly increased, and the emissions from road and transportation facility land are reduced. Figure 8.10 shows the low-carbon measurement results for scheme 2. Compared with scheme 1, the ecological carbon sequestration degree and mixed land-use degree indicators of scheme 2 are all reduced. This result is mainly attributable to the increase in industrial land use and the decrease in green space and plaza land use, roads, and transportation infrastructure.

Based on scheme 1, a second adjustment to land structure and layout was carried out to generate scheme 3. Scheme 3 mainly adjusts the land structure of Fuzhou New Town as follows: increase the proportion of green space by 5%, and reduce the proportion of industrial land by 2%. Table 8.5 shows the adjusted built-up land balance table. The plan after adjustment was reentered into the system for carbon-emission prediction and low-carbon degree evaluation. Figure 8.11 shows the carbon-emission prediction results for scheme 3. The total carbon emissions are 66,336,000 tons/year, the per capita carbon emissions are 30.15 tons/(person*year), and the land average carbon emissions are 131,300 tons/(km^2*year). Compared with scheme 1, the carbon emissions of industrial land in scheme 2 are significantly reduced, and the carbon emissions of road and transportation facility land are slightly reduced. Figure 8.12 shows the low-carbon degree results for scheme 3. Compared with scheme 1, the ecological carbon sequestration degree indicator is increased.

Based on the above test results, the system was able to predict the total carbon emissions and low-carbon measurement of the New Towns' planning scheme based on land use. Along with changes in land-use structure and layout in the planning scheme, the system could also change the total carbon emissions and low-carbon measurement of the scheme accordingly. This is in line with the logical correspondence between land use and carbon emissions as described in Sect. 8.2.4.2 and sensitively reflects the impact of land-use structure and layout changes on the prediction results. This system can effectively estimate total carbon emissions and

Table 8.5 Urban built-up land balance sheet in scheme 3

Land use code	Land use type	Land area (km^2)	Proportion of built-up land (%)
R	Residential	105.02	20.78
A	Administration and public services	19.67	3.89
B	Commercial and business facilities	37.48	7.42
M	Industrial	88.39	17.49
S	Street and transportation	122.96	24.34
W	Logistics and warehouse	23.94	4.74
G	Green space	104.95	20.77
U	Municipal utilties	2.85	0.56
H11	Urban built-up land	505.26	100.00

Fig. 8.11 Carbon-emission prediction results for scheme 3

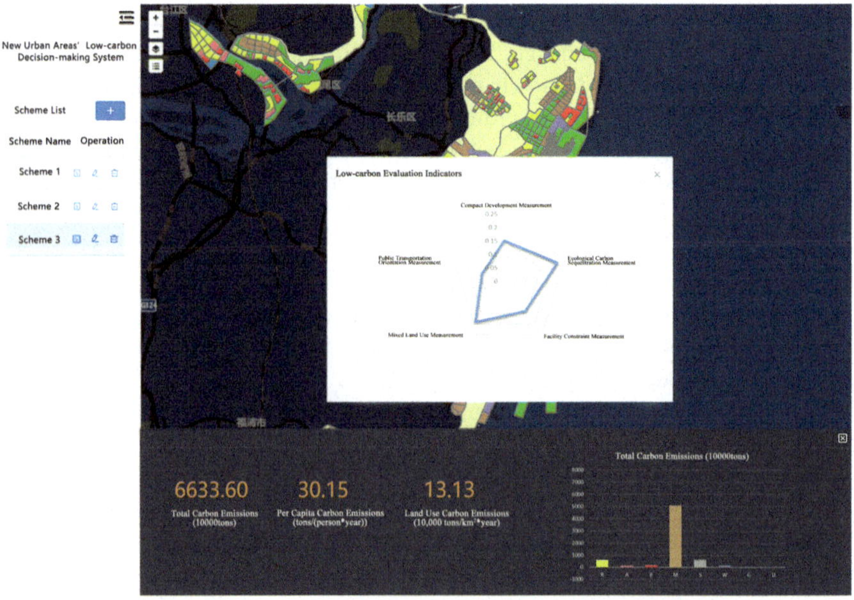

Fig. 8.12 Low-carbon degree measurement results for scheme 3

measure the carbonization of the New Towns' master-plan scheme. Further, it can generate various planning policy scenarios based on different land structures and layout scenarios and analyze total emissions, thus providing decision-making support.

(2) Sensitivity and validity test of the system for the public transportation sharing rate.

Under certain land-use layouts, the use of public transportation will affect transportation emissions and thus affect the city's total carbon emissions. In addition, in the New Towns' master planning stage, the land-use layout often does not involve the planning of public transportation stations, although the public transportation system is closely related to emissions. Therefore, in this study, we designed the public transportation sharing rate as a key global variable affecting emissions in the New Town and tested its sensitivity and validity. As in the previous section, we took Fuzhou New Town as the example. The sensitivity and validity of the system for the public transportation sharing rate was tested by changing the public transportation sharing rate of Fuzhou New Town while the land structure and land layout remained unchanged.

The carbon-emission prediction results for the initial planning scheme (scheme 1 above) and the low-carbon measurement evaluation results remained unchanged (Figs. 8.7, 8.8). Based on scheme 1, when the global variables were input manually, we adjusted the public transportation sharing rate from 40% to 52% and kept other global variables unchanged (Fig. 8.13). At the same time, without adjusting the land

Fig. 8.13 Human–computer interaction: adjusting the key global variables of the planning scheme

structure and layout of the planning scheme, the scheme was input into the system to generate scheme 4.

Figure 8.14 shows the carbon-emission prediction results for scheme 4. The total carbon emissions are 71,530,300 tons/year, the per capita carbon emissions are 32.51 tons/(person*year), and the land average carbon emissions are 141,600 tons/(km^2*year). Compared with scheme 1, only the carbon emissions of road and transportation facility land are slightly reduced. Figure 8.15 shows the Low-carbon degree measurement results for scheme 4. Compared with scheme 1, the public transportation orientation measurement indicator of scheme 4 is increased, indicating that by changing the key global variable of the planning scheme – namely, the public transportation sharing rate – the prediction results for total carbon emissions and low-carbon measurement change accordingly. This change is manifested in the increase in the public transport use rate, which is expected to reduce emissions from intracity traffic, thus improving the public transportation orientation measurement results of the planning scheme.

The test results indicate that the system can calculate traffic emissions based on the public transportation sharing rate in the planning scheme and then predict the total carbon emissions of the New Town. With changes in the public transportation sharing rate, the system's total carbon emissions and the low-carbon measurement of the planning scheme can correspondingly undergo changes in theoretical and mathematical logic. For example, if the transportation use rate is increased without changing other global variables, the land structure, or the planning scheme layout,

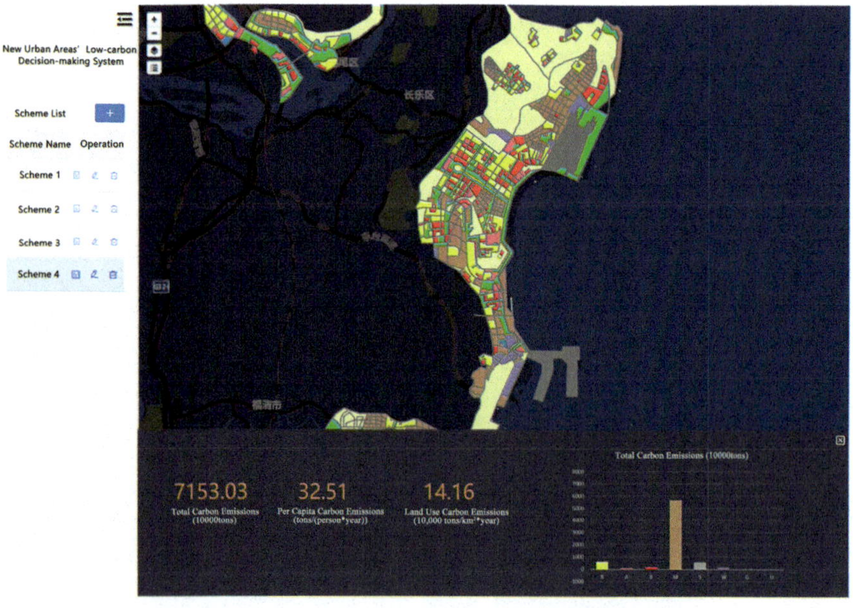

Fig. 8.14 Carbon-emission prediction results for scheme 4

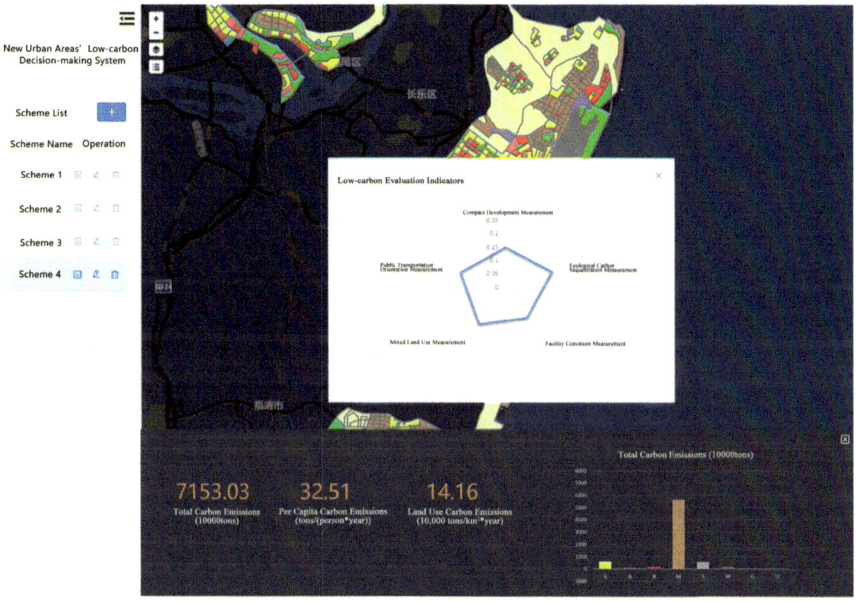

Fig. 8.15 Low-carbon degree measurement results for scheme 4

emissions from roads and transportation facilities will decrease accordingly, and the transportation orientation measurement will increase accordingly. The system can therefore support total emission estimation and low-carbon assessment of New Towns' master-plan schemes based on adjusting key global variables, thus providing support for master-plan decision-making.

8.4 System Application

8.4.1 Overview of the Xixian New Town Master Plan of Xi'an City, China

In 2020, Xixian New Town had a planned population of 2.72 million, urban built-up land of 537.47 km^2, per capita built-up land area of 197.60 m^2, and a gross population density of 424 people/ha (resident population per 1 ha of residential land is 424 people). Figure 8.16 shows the master land-use plan for Xixian New Town.

Xixian New Town is planned to have five clusters: Konggang New Town, Jinghe New Town, Qinhan New Town, Fengdong New Town, and Fengxi New Town (Fig. 8.17). Public facilities will be deployed on the north bank of the Weihe River and the west side of the Fenghe River. Functional public facilities will be reasonably arranged in each cluster, and industrial land will be laid out on both sides of the Jinghe River, the east side of the airport, and the Fengxi group. Storage land will be

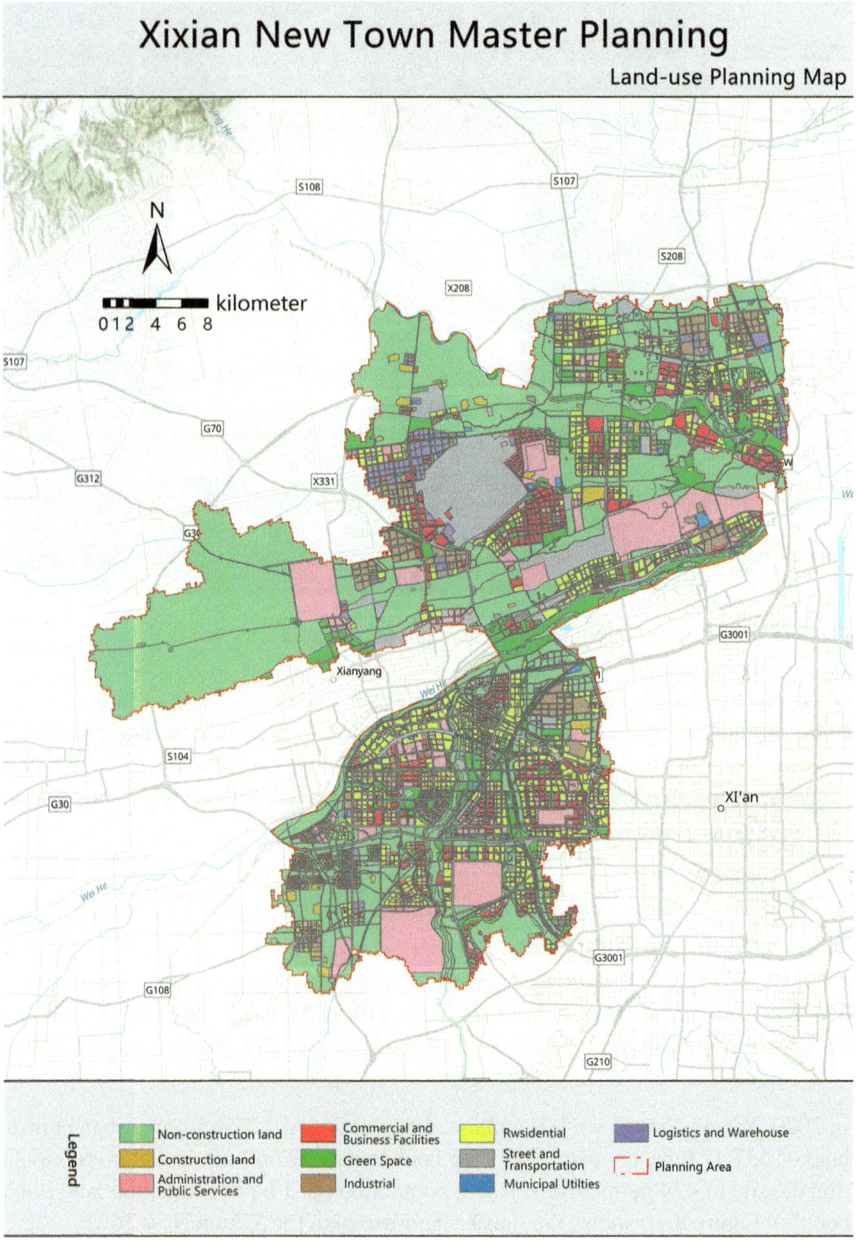

Fig. 8.16 Land-use planning map of the Xixian New Town. (Source: Department of Natural Resources of Shaanxi Province)

arranged on the west side of the airport and the east side of Jinghe New Town. A spatial pattern of "corridor penetration and group layout" will be formed through two imperial tomb relics; the three ecological landscape corridors of Wei River, Feng River, and Jing River; and wedge-shaped green spaces between the clusters.

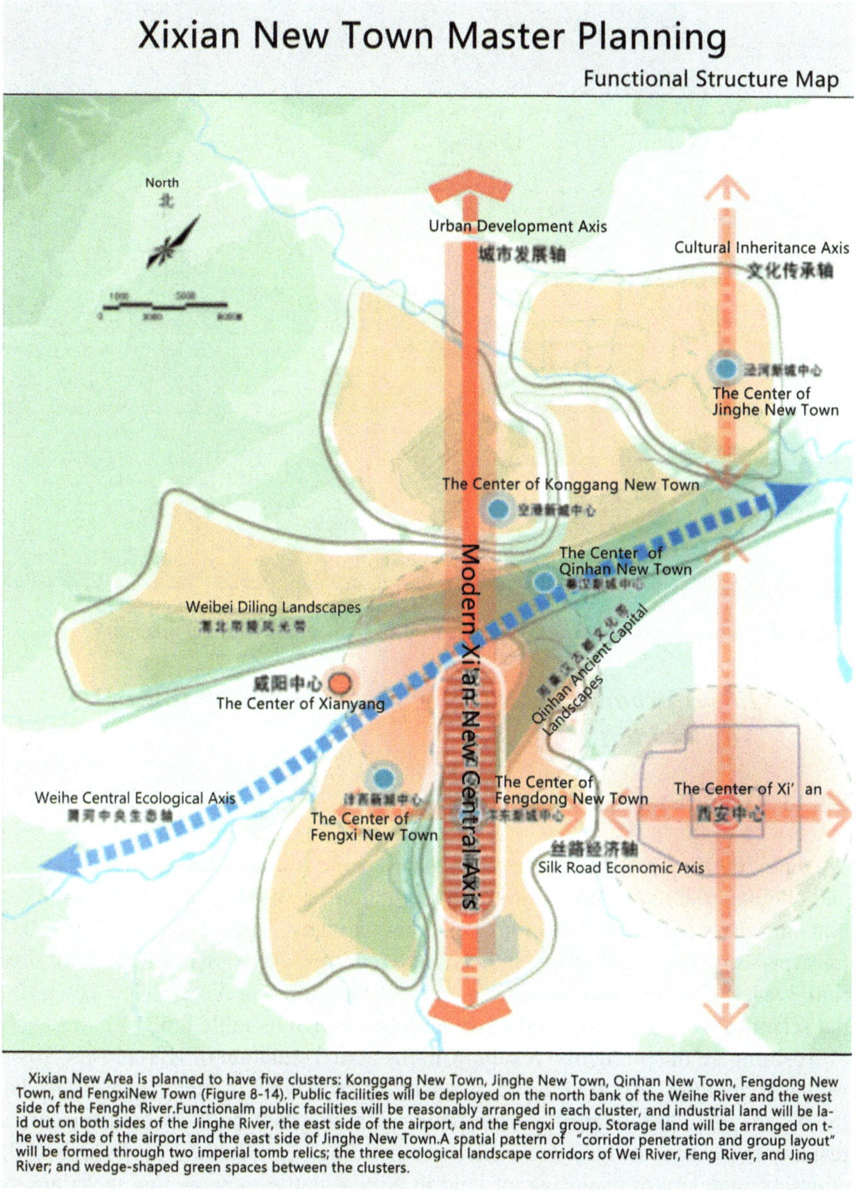

Fig. 8.17 Functional structure map of Xixian New Town. (Source: Department of Natural Resources of Shaanxi Province)

Fig. 8.18 Carbon-emission prediction results for Xixian New Town (initial scheme)

8.4.2 *Low-Carbon Assessment of the Xixian New Town Master Plan*

8.4.2.1 Forecast of Total Carbon Emissions in Xixian New Town

We input the Xixian New Town master-plan scheme (initial scheme) into the system and obtained the total carbon-emission prediction results (Fig. 8.18). The total carbon emissions are 27,302,900 tons/year, the per capita carbon emissions are 10.04 tons/(person*year), and the land average carbon emissions are 54,600 tons/(km^2*year). The prediction results for carbon emissions for Xixian New Town (initial scheme) were then exported into a table, as shown in Table 8.6.

To compare the prediction results with the actual situation of Xixian New Town, we calculated the carbon-emission intensity of each type of land according to data for Xi'an, including energy consumption, area of urban built-up land types, and resident population of the city in the corresponding year (Table 8.7). Assuming the emission intensity of each type of land in Xixian is the same as that in Xi'an, the emission intensity of each land type was first calculated based on Table 8.7, and then the total emissions of Xixian New Town (initial scheme) were calculated as shown in Table 8.8. Comparing emissions calculated based on the land area of roads and transportation facilities in Table 8.8 with emissions calculated automatically by the system based on traffic behavior in Table 8.6, we find large difference between the two results. There are two main reasons for this. First, traffic emissions calculated by the system only included emissions generated by intracity traffic within the planning area of Xixian New Town. Second, the system's traffic-emission prediction is based on the emission-measurement method of traffic behavior. It should be noted that residents' daily travel activities might involve combinations of multiple modes of transport. This method, however, only uses the most preferred mode of transportation travel to finalize travel activity, ignoring the influence of other travel modes. For example, if a resident chooses rail transit and taxi for a trip, it is

Table 8.6 Carbon emissions derived from the system based on traffic behavior in Xixian New Town (initial scheme)

Land use type in Xixian New Town		Carbon dioxide emissions (10,000 tons)
Residential		361.95
Industrial		1623.69
Land for commercial and business facilities, administration and public services and municipal utilties	Commercial and business facilities	198.92
	Administration and public services	377.24
	Municipal utilties	18.22
Land for street and transportation, logistics and warehouse	Street and transportation	99.33
	Logistics and warehouse	56.63
Green space		−5.69

Table 8.7 Carbon-emission intensity table of various land types in Xi'an

Urban land use types		2017 land area in Xi'an (km^2)	2017 Energy Consumption in Xi'an (10,000 tons of standard coal)	Carbon dioxide emissions (10,000 tons)	Land use carbon emission intensity (10,000 tons/km^2)
Residential		144.32	332.04	815.70	5.65
Industrial		78.42	1831.67	4499.80	57.38
Land for commercial and business facilities, administration and public services and municipal utilties	Commercial and business facilities	145.80	241.97	594.44	4.08
	Administration and public services				
	Municipal Utilties				
Land for street and transportation, logistics and warehouse	Street and transportation	127.97	258.85	635.91	4.97
	Logistics and warehouse				
Green space		–	–	–	−0.06

Table 8.8 Carbon emissions calculated based on the carbon-emission intensity of various land types in Xixian New Town (initial scheme)

Urban land use types in Xixian New Town		Land area (km^2)		Land use carbon emission intensity (10,000 tons/ km^2)	Carbon dioxide emissions (10,000 tons)	
Residential		64.04		5.65	361.95	
Industrial		28.30		57.38	1623.69	
Land for commercial and business facilities, administration and public services and municipal utilties	Commercial and business facilities	145.79	48.79	4.08	594.38	198.92
	Administration and public services		92.53			377.24
	Municipal utilties		4.47			18.22
Land for street and transportation, logistics and warehouse	Street and transportation	169.00	157.60	4.97	839.78	783.15
	Logistics and warehouse		11.40			56.63
Green space		93.33		−0.06	−5.69	

identified as rail transit in the system's calculation, which can make the calculation result smaller than the actual value (Zhang 2019).

8.4.2.2 Low-Carbon Measurement of Xixian New Town

We entered the Xixian New Town master-plan scheme (initial scheme) into the system; Fig. 8.19 shows the low-carbon measurement evaluation. The Xixian New Town master-plan scheme (initial scheme) has low indicators in four aspects: compact development measurement, mixed land-use measurement, ecological carbon sequestration measurement, and public transportation orientation measurement. Therefore, those four aspects are the main directions for adjusting the low carbonization of the Xixian New Town master plan.

8.4.3 Multiscenario Regulation of Xixian New Town Master Plan

8.4.3.1 Multipolicy Scenario Design

The abovementioned results for the Xixian New Town master-plan scheme (initial scheme) reveal that the scheme has better indicators in terms of transition orientation measurement, mixed land-use measurement, ecological carbon sequestration measurement, and compact development measurement. Those are the main

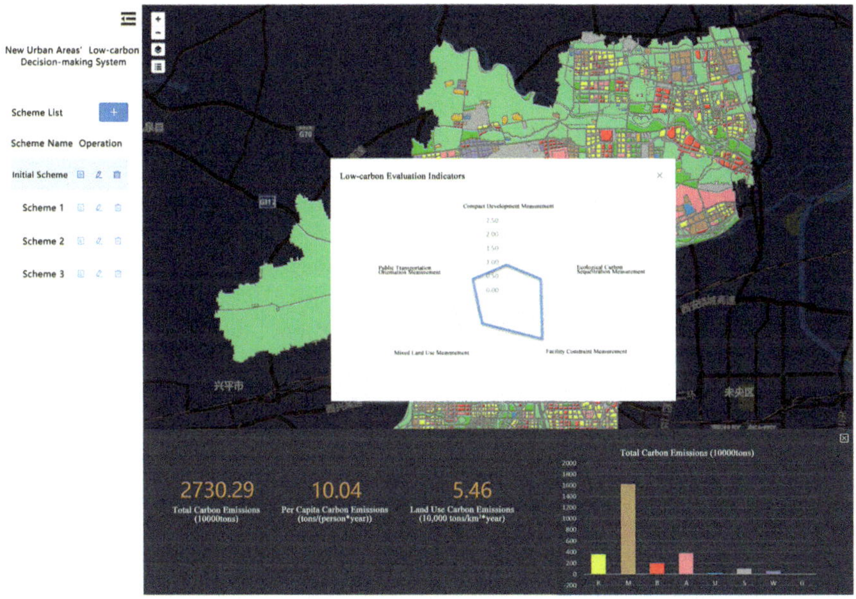

Fig. 8.19 Low-carbon degree measurement results for Xixian New Town master plan (initial scheme)

controlling factors for the low carbonization of the master plan at this stage. Regarding those four aspects, Shaanxi Province and its capital Xi'an have issued various policies, proposing to increase the use rate of public transportation while also focusing on mixed land use and the growth of the "three greens" (green space per capita, green space rate, and green coverage rate). Accordingly, to conduct multipolicy scenario planning, this study set three planning policy scenarios: increasing the public transportation sharing rate, increasing the mixed land use of planning clusters, and increasing the proportion of green space in parks.

8.4.3.2 Policy Scenario 1: Optimize the Transportation Structure and Increase Public Transportation Use

The measurement results for the Xixian New Town master plan revealed that the indicators of public transportation orientation were low. It is known that when the public transportation sharing rate increases, emissions decrease to a certain extent. Therefore, we established policy scenario 1 – namely, optimizing the transportation structure and increasing the public transportation sharing rate – as one of scenario for regulating carbon emissions in the New Town.

Policy scenario 1 includes the following steps:

(1) Increase the public transportation sharing rate

Assuming that bus coverage in Xixian New Town has reached the planning requirements (100% coverage of 300–500 m for general bus stations and 100% coverage of 800–1200 m for rapid rail transit stations) and that residents are willing to accept public transportation as their travel preference, increasing the public transportation sharing rate is conducive to reducing emissions from urban transportation.

According to the *2018 Xi'an Transportation Development Annual Report*, Xi'an's public transportation sharing rate is 35%. Since the initial scheme assumes the public transportation rate in Xixian New Town is at the same level as in Xi'an, the use rate was adjusted in scenario 1 as follows: the public transportation sharing rate was increased from 35% to 45%, with the bus-use rate increased from 24% to 30% and the rail transportation use rate increased from 11% to 15%.

(2) Carbon-emission prediction and low-carbon measurement for the adjusted scheme

Based on the initial planning scheme of Xixian New Town, the public transportation sharing rate was increased to 45%. Subsequently, Xixian New Town planning scheme 1 was input into the system for the carbon-emission prediction and low-carbon measurement evaluation of policy scenario 1. Figures 8.20 and 8.21 show the results.

The results indicate that the total carbon emissions of Xixian New Town planning scheme 1 are 26,951,100 tons, which is a reduction of 351,800 tons. Since there was no change in the land structure and layout, the reduction in total emissions comes from the reduction in road and transportation facility emissions caused by the increase in public transportation use. Per capita carbon emissions are reduced by 0.07 tons/(person*year) and by 1300 tons/(km^2*year). At the same time, the bus

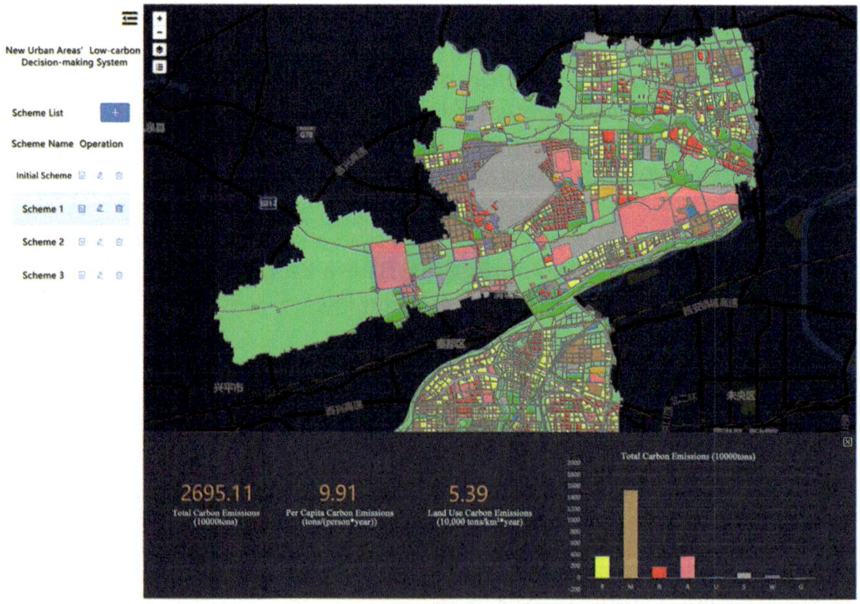

Fig. 8.20 Carbon-emission prediction results for Xixian New Town planning scheme 1

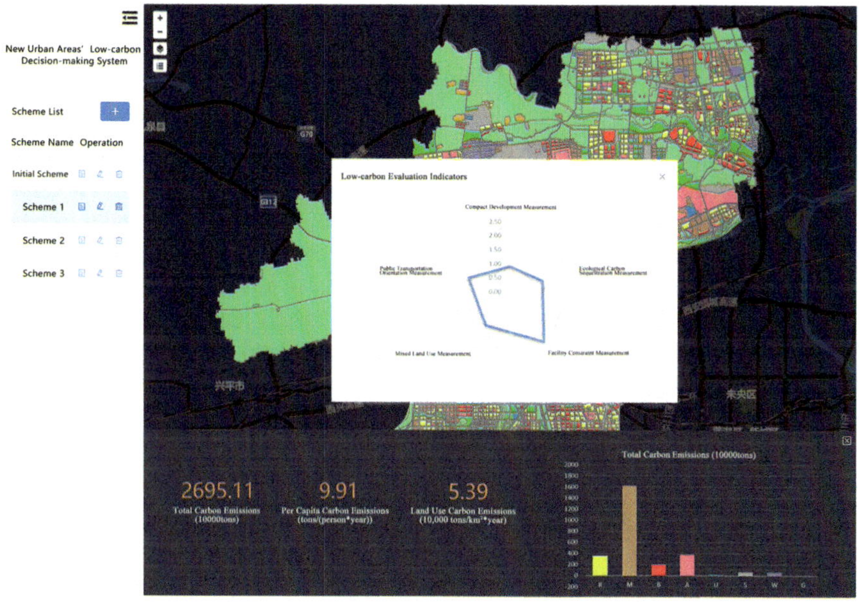

Fig. 8.21 Low-carbon degree measurement results for Xixian New Town planning scheme 1

orientation measurement of Xixian New Town planning scheme 1 is also improved from 1.3 to 1.5; that is, when the bus-use rate is increased by 1%, the bus orientation measurement correspondingly increases by 1.33%.

Some studies have shown that the per capita carbon-emission factor of small cars is 2.4 times that of buses and 19.6 times that of rail transit for the same travel distance (Zhang and Jia 2020). When the urban transportation structure is optimized and the urban public transportation sharing rate increases, more residents will choose public transportation and reduce their individual car trips, thereby reducing emissions. Thus, the policy can help reduce emissions in Xixian New Town and improve the indicators of public transportation orientation measurement. According to the actual situation in Xixian New Town, for every 1% increase in public transportation use, total emissions will be reduced by 35,200 tons, and the public transportation orientation measurement will increase by 1.33%.

8.4.3.3 Policy Scenario 2: Improve the Degree of Mixed Land Use in Monofunctional Clusters

The Low-carbon degree measurement results for the Xixian New Town master plan indicate that the indicators of mixed land-use measurement are low. Studies have shown that a higher degree of mixed land use can shorten travel distances for residents (Zhu and Niu 2020), thereby further reducing traffic emissions. Thus, we established policy scenario 2 – that is, improving the degree of mixed land use in more monofunctional clusters – as another way of regulating emissions in the New Town.

Policy scenario 2 includes the following steps:

(1) Measure the degree of mixed land use in Xixian New Town (initial scheme)

According to the land-use planning map of Xixian New Town (initial scheme), the spacing between most of branch roads is 400–500 m. Therefore, from the perspective of ensuring the integrity of neighborhoods and the operability of planning, a 500*500 m grid was selected as the basic spatial unit for mixed land-use assessment to measure the degree of mixed land use. Information entropy was used to calculate the degree of mixed land use. The closer the entropy value is to 0, the more homogeneous the land use within the unit; the larger the entropy value, the higher the degree of mixed land use. The formula is as follows:

$$H = -\sum_{i=1}^{n} p_i \cdot \log_2 (p_i) \tag{8.14}$$

where p_i is the probability of occurrence of a certain type of land that affects residents' travel. In this study, the area of a certain type of land divided by the total area of the basic unit represents the probability of occurrence of this type of land. H represents the degree of mixed land use.

We imported the land-use planning map of Xixian New Town into ArcGIS to calculate the degree of mixed land use; Fig. 8.22 shows the results.

The measurement results were classified using the natural break point method. The basic spatial unit entropy value $H \leq 0.7$ indicates a low degree of mixed land use, $0.7 < H \leq 1.0$ indicates a medium degree of mixed land use, and $H \geq 1.0$ indicates a high degree of mixed land use.

Figure 8.22 reveals that the degree of mixed land use in each basic spatial unit is different, and land with a low mixed-use degree is distributed in a plane shape in each cluster. The main reason is that the low mixed-use land in the middle of Qinhan New Town and the south of Fengdong and Fengxi New Town is mostly located in heritage protection areas and construction control zones. Further, the low mixed-use land in the middle of Konggang New Town is the site of Xi'an Xianyang International Airport. Thus, the low mixed-use land in the above two cases were excluded from subsequent adjustments. The high mixed-use land is distributed in the form of large scattering and small clustering. The basic spatial units are mostly composed of residential land and mixed-use land for commercial services, public management, and public facilities. Among them, the clustering of high mixed-use land in Fengdong and central Fengxi is more obvious.

(2) Identify the cluster with the lowest degree of mixed land use in Xixian New Town

The degree of mixed land use in Xixian's five major clusters was calculated again in ArcGIS using the same formula as above; Fig. 8.23 shows the results.

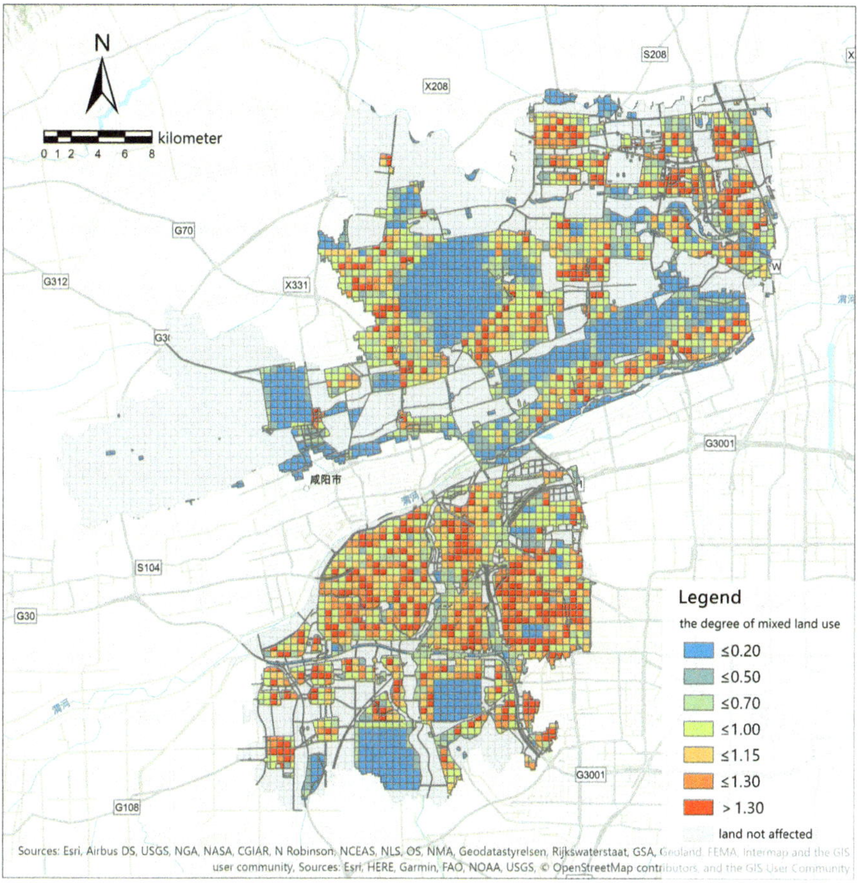

Fig. 8.22 Degree of mixed land-use measurement results for Xixian New Town based on spatial grid cells

Degrees of mixed land use in the clusters were identified as follows (highest to lowest): Jinghe New Town cluster (mixed degree: 1.82), Fenxi New Town cluster (1.71), Fengxi New Town cluster (1.70), Qinhan New Town cluster (1.66), and Konggang New Town cluster (1.57). Thus, the cluster with the lowest mixed land-use degree was the Konggang New Town cluster.

(3) Adjust land use in the Konggang New Town cluster

The land-use structure and layout of the Konggang New Town cluster were manually adjusted to improve the degree of mixed land use. Figure 8.24 shows the adjusted land-use planning map of Xixian New Town planning scheme 2, and Table 8.9 shows the land-use structure.

Compared with the initial scheme, the following changes are observed in the land-use structure of Xixian New Town planning scheme 2: the proportion of residential land use increased by 0.57%, the proportion of green space increased

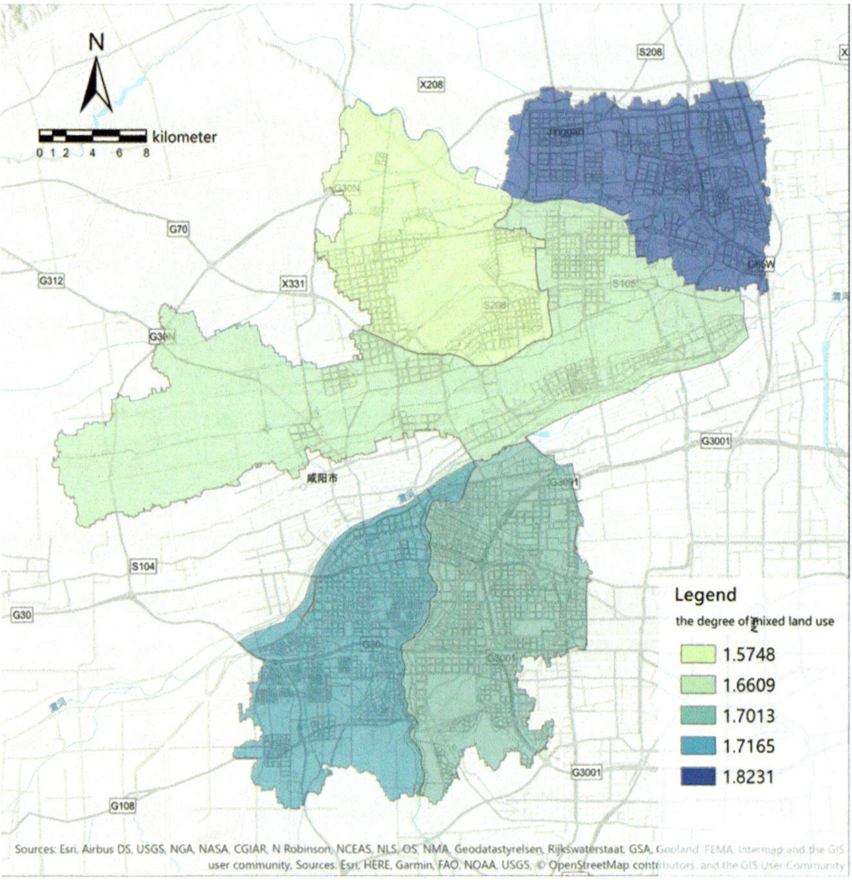

Fig. 8.23 Degree of mixed land-use measurement results for five major clusters in Xixian New Town

by 0.23%, the proportion of commercial service facilities decreased by 0.22%, the proportion of industrial land use decreased by 0.32%, and the proportion of logistics and warehouse land use decreased by 0.25%. Since the degree of mixed land use was changed mainly by adjusting the land-use layout, the change in the control land structure was relatively small.

After making the abovementioned adjustments, the degree of mixed land use was measured again using the same method; Fig. 8.23 shows the results. Figures 8.25 and 8.26 show the compared results for the degree of mixed land use in Konggang New Town before and after the adjustment. Tables 8.10 and 8.11 show the statistical output for the degree of mixed land use. The proportion of basic cell grids with a high degree of mixed use is increased while the proportions with medium and low degrees are both reduced. This indicates that the degree of mixed land use in Konggang New Town was effectively improved after the adjustment.

(4) Carbon-emission prediction and low-carbon measurement for the adjusted scheme

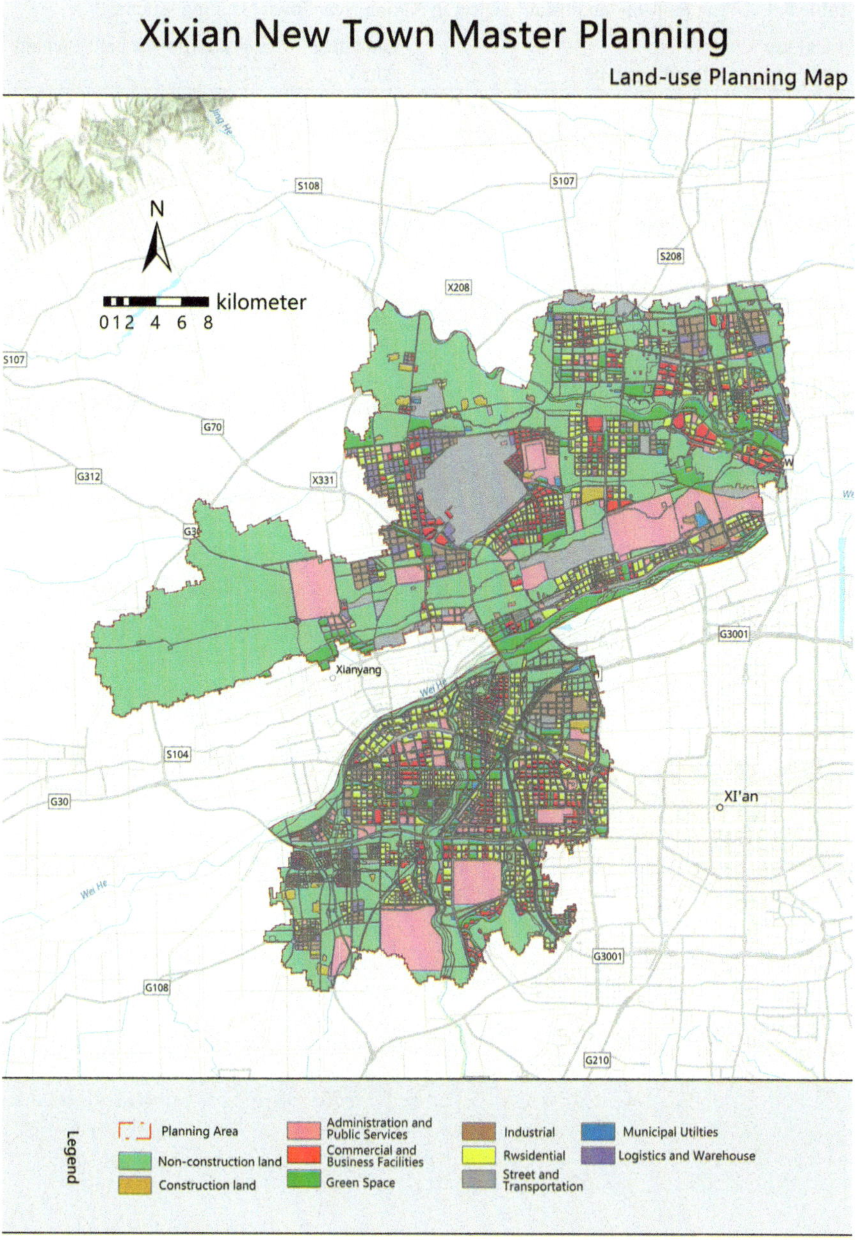

Fig. 8.24 Land-use planning map in Xixian New Town planning scheme 2

We input the adjusted Xixian New Town planning scheme 2 into the system to predict the carbon-emission and evaluate the low-carbon degree. Figures 8.27 and 8.28 show the results. After the adjustment, the total carbon emissions of scheme 2 are 26.4221 million tons – a reduction of 880,800 tons. The changes in carbon emissions due to changes in land structure are as follows: residential

Table 8.9 Urban built-up land balance sheet in Xixian New Town planning scheme 2

Land use code	Land use type	Land area (km^2)	Proportion of built-up land (%)
R	Residential	66.87	13.36
A	Administration and public services	92.53	18.49
B	Commercial and business facilities	47.68	9.53
M	Industrial	26.69	5.33
S	Street and transportation	157.60	31.49
W	Logistics and warehouse	10.13	2.02
G	Green space	94.49	18.88
U	Municipal utilties	4.47	0.89
H11	Urban built-up land	500.46	100.00

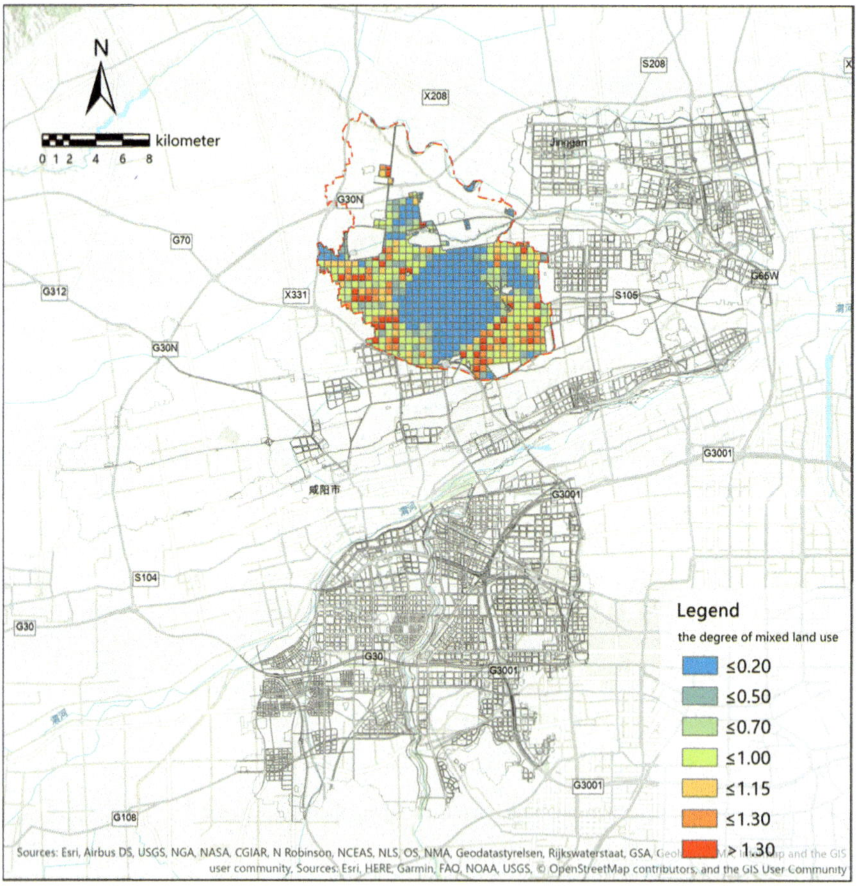

Fig. 8.25 Degree of mixed land-use measurement results for Konggang New Town before adjustment

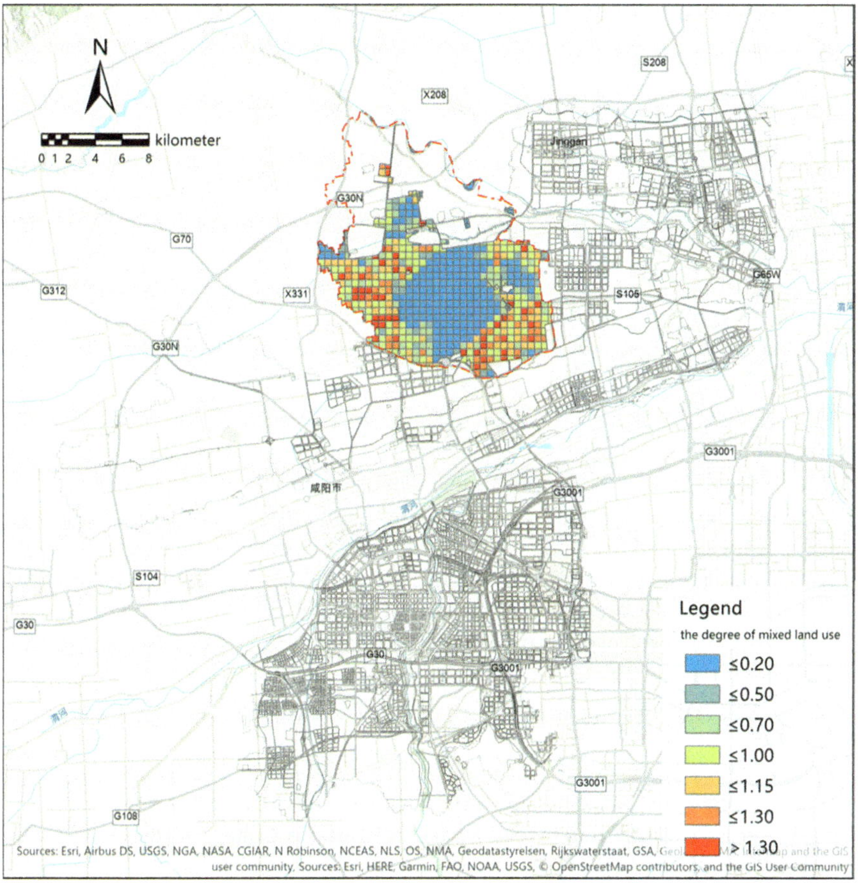

Fig. 8.26 Degree of mixed land-use measurement results for Konggang New Town after adjustment

land carbon emissions increased by 159,900 tons; industrial land carbon emissions decreased by 922,400 tons; commercial services, public management, and public facilities emissions decreased by 46,100 tons; logistics and warehouse land emissions were reduced by 62,900 tons; and the carbon sink generated by green space increased by 700 tons. Regarding changes in emissions due to changes in mixed land use, emissions from roads and transportation facilities were reduced by 8600 tons.

Accounting for the actual situation of Xixian New Town, we find that when the land structure changes, the proportion of industrial land decreases by 1%, and emissions from industrial land decrease by 2,882,500 tons. When the proportion of commercial services, public administration, and public facilities land decreases by 1%, emissions decrease by 209,500 tons. When the proportion of logistics and warehouse land decreases by 1%, emissions decrease by 251,600 tons. When the degree of mixed land use changes, for every 1% increase, emissions from road and transportation facilities are reduced by 39,100 tons.

Table 8.10 Statistical table of degree of mixed land use in Konggang New Town

Mixed degree	Number of basic space units	Proportion (%)
Low ($0 < H \leq 0.7$)	415	60.67
Medium ($0.7 < H \leq 1.0$)	130	19.01
High ($H > 1.0$)	139	20.32

Table 8.11 Statistical table of degree of mixed land use in Konggang New Town after adjustment

Mixed degree	Number of basic space units	Proportion (%)
Low ($0 < H \leq 0.7$)	342	55.34
Medium ($0.7 < H \leq 1.0$)	116	18.77
High ($H > 1.0$)	160	25.89

Figure 8.28 shows the Low-carbon degree measurement results for Xixian New Town planning scheme 2. The degree of mixed land use improved from 1.5 to 1.6. It can be seen from the results that when the degree of mixed land use increases by 0.22%, the measurement increases by 6.6%; that is, for every 1% increase in the degree of mixed land use, the mixed land-use measurement increases by 30%. It is clear, then, that this policy can help reduce the total emissions of Xixian New Town and improve the measurement indicator of mixed land use. Based on the actual situation of Xixian New Town, every 1% increase in the degree of mixed land use will reduce emissions by 39,100 tons, and the measurement of mixed land use increases by 30% correspondingly.

8.4.3.4 Policy Scenario 3: Adjust the Land Structure and Increase the Proportion of Green Space in Built-Up Land

The measurement results for the Xixian New Town master plan revealed that the indicators of ecological carbon sequestration measurement were low. Thus, we established policy scenario 3 – that is, adjusting the land structure and increasing the proportion of green space in built-up land – as one of the scenarios for regulating emissions in the New Town.

Policy scenario 3 includes the following steps:

(1) Increase the proportion of green space by 5%, and conduct carbon-emission prediction and low-carbon measurement assessment again

Figures 8.18 and 8.19 show that the carbon-emission prediction results and low-carbon measurement assessment of Xixian New Town (initial scheme) remain unchanged. Table 8.12 shows the built-up land balance table of Xixian New Town (initial scheme).

Based on the initial planning scheme for Xixian New Town, the land structure and layout were adjusted as follows: the proportion of green space was increased by 5% and that of residential, administration and public services, commercial and business facilities, industrial, street and transportation was decreased by 1% each. Table 8.13 shows the adjusted built-up land balance

Fig. 8.27 Carbon-emission prediction results for Xixian New Town planning scheme 2

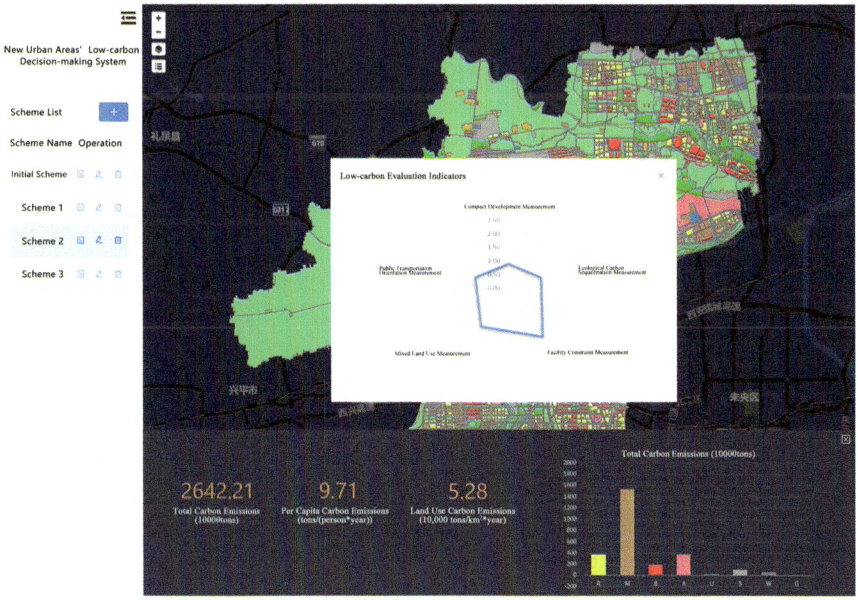

Fig. 8.28 Low-carbon degree measurement results for Xixian New Town planning scheme 2

table. Subsequently, Xixian New Town planning scheme 3 was input into the system for carbon-emission prediction and low-carbon measurement evaluation; Figs. 8.29, 8.30 show the results.

The results show that the total carbon emissions of Xixian New Town planning scheme 3 are 23,723,900 tons, for a total reduction of 3,579,000 tons. Specifically, emissions from residential land decrease by 283,100 tons, emissions from land for public management and public services decrease 204,200 tons, emissions from land for commercial service facilities decrease by 203,800 tons, industrial land emissions decrease by 2,872,900 tons, and green-space carbon sinks increase by 15,300 tons. Per capita carbon emissions are reduced by 0.72 tons/(person*year) and by 13,200 tons/(km^2*year).

Compared with the original scheme, the ecological carbon sequestration measurement of Xixian New Town planning scheme 3 increases from 1.3 to 1.7, and the mixed land-use measurement increases from 1.5 to 1.75. That is, when the proportion of green space increases by 1%, the ecological carbon sequestration degree will increase by 4.62%, and the degree of mixed land use increases by 2.67% accordingly. The results indicate that changing the land structure and layout of the planning scheme can help reduce the total emissions of Xixian New Town and improve the ecological carbon sequestration measurement and mixed land-use measurement indicators. Based on the practical situation of Xixian New Town, when the proportion of green space increases from 18.65% to 23.65%, for every 1% increase in the proportion of green space, total carbon emissions are reduced by 715,800 tons, the ecological carbon sequestration increases by 4.62%, and the degree of mixed land use increases by 2.67%.

(2) Adjust the proportion of green space to increase by 7% and conduct carbon-emission prediction and low-carbon measurement assessment again

The land structure and layout of Xixian New Town were adjusted again as follows: the proportion of green space was increased by 7%; that of logistics and warehouse land was decreased by 0.5%; that of residential land, land for

Table 8.12 Urban built-up land balance sheet in the initial scheme

Land use code	Land use type	Land area (km^2)	Proportion of built-up land (%)
R	Residential	64.04	12.80
A	Administration and public services	92.53	18.49
B	Commercial and business facilities	48.79	9.75
M	Industrial	28.30	5.65
S	Street and transportation	157.60	31.49
W	Logistics and warehouse	11.40	2.28
G	Green space	93.33	18.65
U	Municipal utilties	4.47	0.89
H11	Urban built-up land	500.44	100.00

Table 8.13 Urban built-up land balance sheet in Xixian New Town planning scheme 3

Land use code	Land use type	Land area (km^2)	Proportion of built-up land (%)
R	Residential	59.03	11.80
A	Administration and public services	87.52	17.49
B	Commercial and business facilities	43.79	8.75
M	Industrial	23.29	4.65
S	Street and transportation	152.60	30.49
W	Logistics and warehouse	11.40	2.28
G	Green space	118.35	23.65
U	Municipal utilties	4.47	0.89
H11	Urban built-up land	500.44	100.00

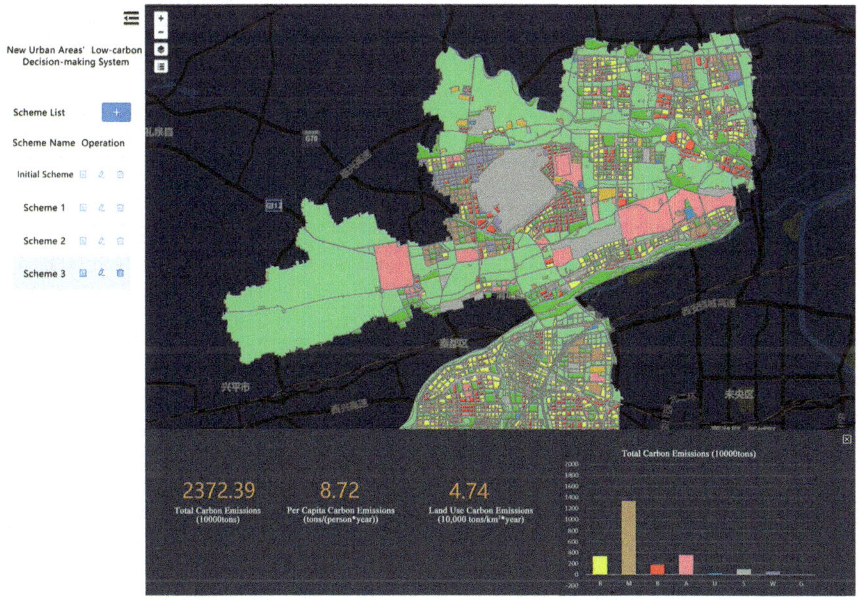

Fig. 8.29 Carbon-emission prediction results for Xixian New Town planning scheme 3

public administration and public services, and industrial land was decreased by 1% each; that of land for road and transportation facilities was decreased by 1.5%; and that of land for commercial service facilities was reduced by 2%. Table 8.14 shows the adjusted built-up land balance table. Xixian New Town planning scheme 4 was then input into the system for carbon-emission prediction and low-carbon measurement evaluation. Figures 8.31 and 8.32 show the results.

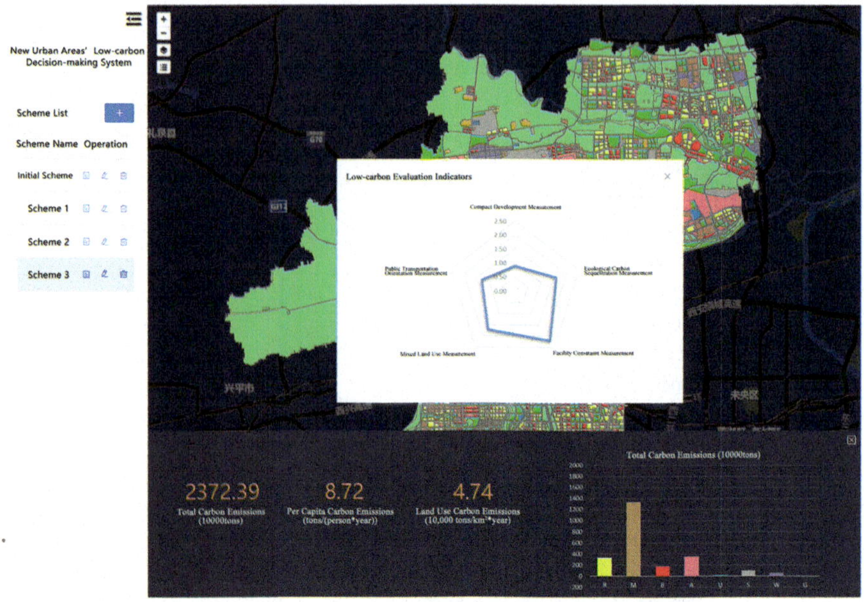

Fig. 8.30 Low-carbon degree measurement results for Xixian New Town planning scheme 3

Table 8.14 Urban built-up land balance sheet in Xixian New Town planning scheme 4

Land use code	Land use type	Land area (km^2)	Proportion of built-up land (%)
R	Residential	59.03	11.80
A	Administration and public services	87.52	17.49
B	Commercial and business facilities	38.78	7.75
M	Industrial	23.29	4.65
S	Street and transportation	150.09	29.99
W	Logistics and warehouse	8.89	1.78
G	Green space	128.36	25.65
U	Municipal utilties	4.47	0.89
H11	Urban built-up land	500.44	100.00

Compared with the original scheme, the total carbon emissions of Xixian New Town planning scheme 4 are 23,388,800 tons, for a total reduction of 3,914,100 tons. Specifically, emissions from residential land decrease by 283,100 tons, those from land for public management and public services decrease by 204,200 tons, those from land for commercial service facilities decrease by 408,100 tons, those from industrial land decrease by 2,872,900 tons, those from land for logistics and storage decrease by 124,500 tons, and those from green space sinks increase by

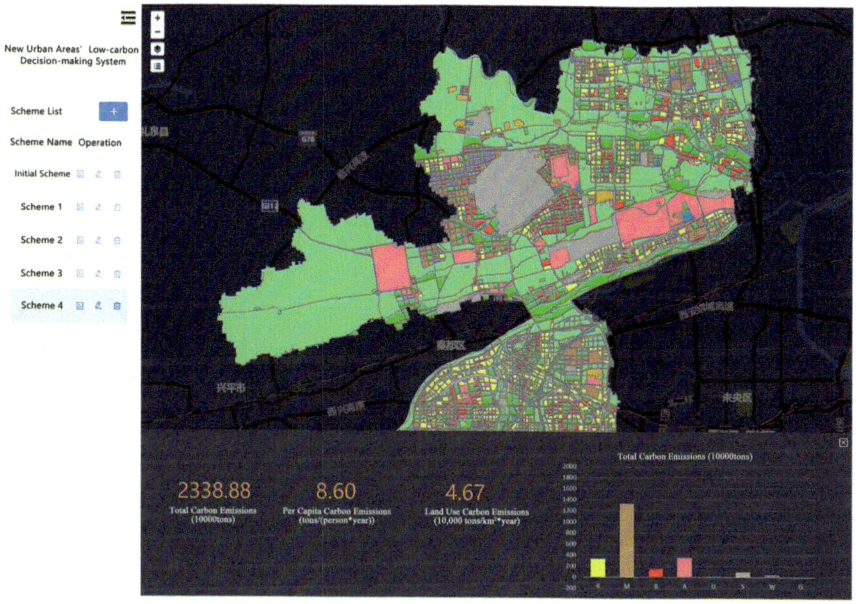

Fig. 8.31 Low-carbon degree measurement results for Xixian New Town planning scheme 3 after policy scenario 3 intervention

21,400 tons. Per capita emissions are reduced by 0.78 tons/(person*year) and by 14,400 tons/(km^2*year).

Figure 8.32 shows the Low-carbon degree measurement results for scheme 4. Compared with the original scheme, the ecological carbon sequestration measurement increases from 1.3 to 1.7, and the mixed land-use measurement increases from 1.5 to 1.75. That is, when the proportion of green space increases by 1%, the ecological carbon sequestration degree increases by 4.39%, and the degree of mixed land use increases by 2.38%. These results indicate that changing the land structure and layout of the planning scheme can reduce total emissions and improve the ecological carbon sequestration measurement and mixed land-use measurement indicators. According to the practical situation of Xixian New Town, when the proportion of green space increases from 18.65% to 25.65%, for every 1% increase in the proportion of green space, total carbon emissions will be reduced by 559,200 tons, the ecological carbon sequestration will increase by 4.39%, and the degree of mixed land use will increase by 2.38%.

Comparing the prediction of carbon emission and the assessment results of low carbon degree of the two planning schemes, emissions are reduced and the low carbon degree improved as the proportion of green space in urban built-up area increases. However, urban planning must consider the needs of urban development, and the proportion of green space cannot be increased indefinitely. Therefore, urban planners should increase green space to an appropriate proportion according to development needs and innovate the greening mode, such as vertical greening, roof garden, etc.

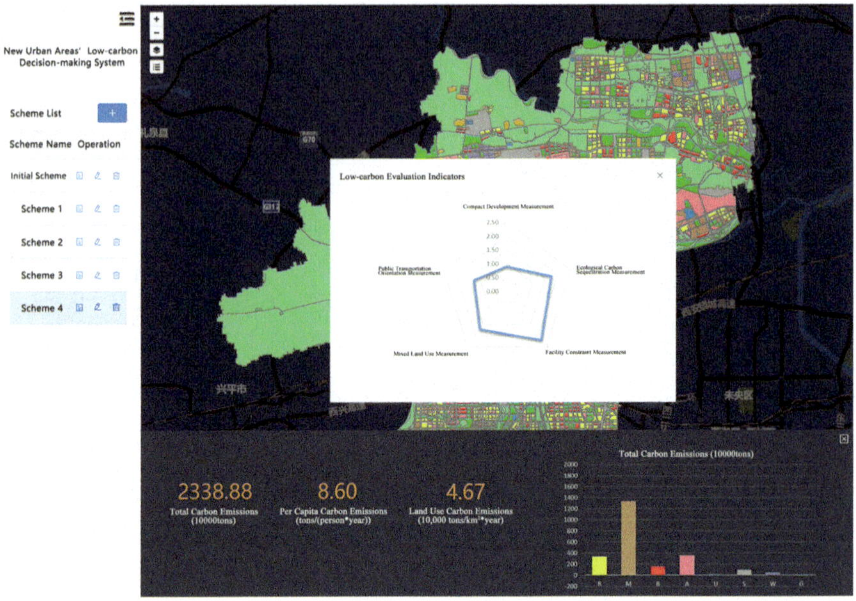

Fig. 8.32 Low-carbon degree measurement results for Xixian New Town planning scheme 4

8.5 Conclusion

This study developed a decision-making system for low-carbon New Towns in China and took Xixian New Town as an example for analysis. First, we performed total carbon-emission prediction and low-carbon degree assessment for the Xixian New Town master plan. Then, on that basis, we explored three policy scenarios that involved increasing the public transportation sharing rate, improving the degree of mixed land use, and increasing the proportion of green space.

Total emissions decreased under all three scenarios, but such reductions significantly differed under different policy scenarios. Specifically, for each 1% increase in the public transportation sharing rate, emissions are reduced by 35,200 tons. For each 1% increase in the degree of mixed land use, emissions are reduced by 39,100 tons. For each 1% increase in the share of green space between 18% and 23%, emissions are reduced by 715,800 tons. Finally, for each 1% increase in the share of green space between 18% and 25%, emissions are reduced by 559,200 tons.

When the planning scheme is adjusted in terms of changing the public transportation sharing rate or the degree of mixed land use, the effect on emissions is mainly reflected in reducing emissions from transportation. Therefore, the numerical change in total carbon emissions is not obvious. When the adjustment changes the land-use structure – that is, the area of each land-use type changes because the emission intensity of each is different – the increase or decrease in emissions caused by different types of land use varies greatly. For example, the emission intensity of industrial land is 573,800 tons/km^2, which is the strongest among the eight land types. As the proportion of industrial land decreases, emissions are more obviously reduced. Aside from green spaces, the emission intensity of other land types ranges from 40,000 to 60,000 tons/km^2. The change in total emissions caused by adjusting the proportion of these land types is not as obvious as in the case of industrial land.

The results of the three policy scenario simulations indicate that the emission intensity of industrial land is the main controlling factor for low carbonization in New Towns, followed by the degree of mixed land use and the public transportation sharing rate. Although parks and green spaces play important roles in carbon sinks, given the characteristics of the land-use layout in the urban master plan, increasing the proportion of those spaces will reduce the area of other land types. Therefore, reducing emissions by increasing the proportion of parks and green spaces is not the optimized solution. This is because that comparing with the emissions produced the industry, the carbon sink produced the green spaces is too less. The energy structure is still the major influencing factor in reducing emission. While in urban master plan, the mixed land use, the proper green space, the high public transportation sharing rate are the key factors influencing on carbon emission.

Achieving the low-carbon goals of New Towns should be a multidimensional intervention based on low-carbon measurements. This cannot be achieved by simply reducing industrial land use or increasing green spaces. In the planning process, under the premise of ensuring a reasonable proportion of parks and green spaces, improving the degree of mixed land use and the public transportation sharing rate will play important roles in emission reduction. Although the proposed system cannot directly help planners generate planning schemes, it can provide a convenient and efficient method for emission prediction and low-carbon assessment for the analysis and comparison of multiple schemes.

References

Zhang LL. Research on Structure Optimization and Regulation of Urban Transportation Travel under the Target of Carbon Emission Reduction[D]. China University of Mining and Technology, 2019. (in Chinese)

Zhu J, Niu XY. Jobs-housing balance, land use mix and relationships with commuting distance:a study of nanning city based on cellphone signaling data [J]. Modern Urban Research, 2020(02):98–105+116. (in Chinese)

Zhang C, Jia FJ. Characteristics of Carbon Emissions from Urban and Strategic of Energy Saving and Emission Reduction Based on Resident Travel[J]. Traffic & Transportation, 2020, 36(03):76–79. (in Chinese)

Index

Y. Ma, Z. Shen, *Strategic Spatial Planning Support System for Sustainable Development*, Advances in Geographic Information Science,
https://doi.org/10.1007/978-3-031-07543-8

The manufacturer's authorised representative in the EU is Springer
Nature Customer Service Centre GmbH, Europaplatz 3, 69115 Heidelberg,
Germany. If you have any concerns regarding our products, please
contact ProductSafety@springernature.com

Printed and bound by CPI Group (UK) Ltd, Croydon, CR0 4YY
01/10/2026
02225872-0001